On every side the sky, on every side the sea

(*Caelum undique et undique pontus*), Virgil, Aeneid, III, 193

ON EVERY
SIDE THE SEA

Man's Involvement
with the Oceans

by ROBERT CLAIBORNE

AMERICAN HERITAGE PRESS • NEW YORK

Other Books
by Robert Claiborne

TIME
CLIMATE, MAN AND HISTORY

CONTENTS

CHAPTER I
THE ENCIRCLING OCEAN

Run your eye over almost any ancient map of the world and you will see the land areas—Europe and a truncated Asia and Africa—huddled about the Mediterranean, which the Romans familiarly called "Our Sea." But encircling the land on every side is The Ocean, endless and so forbiddingly mysterious that not even the arrogant Roman imperialists dared assert any proprietary rights over it.

Modern globes show far larger continents than the Romans dreamed of, and more of them. But although the land has expanded, the sea has expanded faster. We now know that it covers seven-tenths of the earth's surface and is deep enough, if our planet's crust were leveled to a featureless plain, to submerge the entire globe beneath more than two miles of water.

And still it encircles the land. There is no continental mass so huge but that the traveler will not, sooner or later, reach the seashore. Yet at latitude 60 degrees south you can sail eastward, the westerly gales and their house-high billows behind you, with never a landfall. The division of The Ocean into three or four oceans and a score of seas is strictly a geographer's convention; no man can say at what point below Cape Horn the Atlantic leaves off and the Pacific begins, or where either of them becomes the Indian Ocean. The waves that today pound the beaches of Cape Cod or Monterey were once flipped into spray by flying fish south of Java or churned by whales in their feeding grounds off icy Antarctica.

Nor is this encirclement merely a literal truth of geography. Life on land is figuratively encompassed by the sea because it is dependent, indirectly and often directly, too, on the sea for its existence. Without the moisture whose ultimate source is the evaporating oceans, every land-dwelling organism would soon wither away into inanimate dust. Without the shoals of fish that the sea brings forth, hundreds of millions of birds would perish and millions of human beings would sicken for lack of protein. The sea influences the land far more than the land influences the sea; only within the past century or so has man, with more resources in energy and materials than any other creature, managed to significantly alter even the edges of the ocean. We exist by the grace of the sea, and as the shadow of our technological civilization looms ever larger over both land and ocean, this is a fact worth keeping in mind.

The sea's pervasive significance for living organisms goes back to the very beginning of life on our planet. The existence of water in liquid form is an absolute precondition for life as we know it, and perhaps for life of any kind. The only planet, apart from our own, on which we have even slight reason to believe that life exists is Mars—oceanless, but apparently possessing faint traces of liquid water. Significantly, Martian life, so far as we can now tell, appears to be quite as impoverished (if it exists at all) as the planet's water resources.

Water's special hospitality to life is largely due to its capacity for dissolving almost anything. Sea water contains more than two-thirds of the ninety naturally occurring elements in measurable concentrations, and quite probably most of the remainder in infinitesimal amounts; the list of chemical compounds dissolved in the oceans runs well into the hundreds. And living organisms, even the most complex, are fundamentally just sacks full of chemicals whose reactions—most of which can occur *only* in solution—constitute life.

It is hardly surprising, then, that living matter originated in the sea and remained there for most of its time on earth. Life has existed on land for the very respectable period of some 300 million years—but it has populated the sea for approximately ten times as long. Before organisms could make their way onto terra firma, they had to evolve elaborate mechanisms for maintaining the vital water within their tissues. We ourselves are still more than two-thirds water, despite the eons that have passed since our water-dwelling ancestors first hauled themselves onto some prehistoric mud flat. Our bloodstreams still manifest their kinship to the ocean from which we came, for though blood is distinctly less salty than sea water, it maintains almost the same proportions of sodium to potassium and calcium. Whenever we swallow a glass of water, or sprinkle salt on our food, we testify to our marine origins; when a physician infuses water into the veins of a dehydrated patient, he must add salt to it lest he disturb the chemical balance of the tiny ocean within.

A second reason for the sea's long monopoly of life is that it is a relatively undemanding environment. Water, that prime necessity, is always available; moreover, it tempers the fiercest rays of the sun—which can blind or blister when unfiltered. And its buoyancy frees marine organisms from most or all of the labor of maintaining themselves against the ever-present pull of gravity, so that they can conserve their energies for the really important biological tasks of feeding and reproduction.

Finally, the ocean is in one vital respect—temperature—a far more stable environment than land. For reasons deriving from the architecture of the H_2O molecule, water has a higher specific heat than any other common substance. This means that relatively large quantities of heat are required to raise its temperature even a few degrees—a fact that will come as no surprise to anyone who has ever waited for a kettle to boil. In few parts of the ocean surface do temperatures vary more than 10 degrees between summer and winter; day and night variations are much smaller—and a few fathoms below the surface even these differences are largely damped out. By contrast, temperatures on land (except in the equatorial regions) can shift 20 degrees or more within a few hours, while over a year they range, in some places, from well below freezing to more than 100 degrees in the shade. Land organisms have had to work out ways of adapting to these temperature changes—which would be catastrophic for nearly all sea creatures.

A natural consequence of the sea's undemanding nature is that marine organisms are both less diversified and, on the average, simpler in structure than their cousins on land. The great majority of marine plants are one-celled; of the more sophisticated ("vascular") plants, only a handful of species live in the oceans. The same is true of animals, although at first glance marine animals *seem* to be more diversified than land ones. Of the approximately two dozen major groups (phyla) of animals, nearly half are exclusively marine while only one is wholly terrestrial. However, of the wholly sea-dwelling phyla, all but one are small and biologically marginal, comprising at most only a few hundred species (at this time one phylum includes only three species). Nearly all marine animals are rather primitively constructed—endless variations on the theme of how to succeed at living without trying very hard. On the other hand, all three of the biggest and most successful phyla—the mollusks, the arthropods (crabs, spiders, insects, etc.), and our own phylum of the chordates—are well represented on land. The insects alone have more than 700,000 known species—nearly three times as many as all other animals put together—and not one of them lives in the ocean.

The more demanding nature of land living is also shown by the fact that over the past 300 million years or so not a single group of marine animals has managed, so far as we know, to make its way onto the land against the competition of the groups that had already adapted to that strenuous environment. During the same period, on the other hand, scores of land species have gone back to the sea—and have, for the most part, competed very successfully with marine species, despite their persisting handicap of having to obtain their oxygen from the air rather than from the water. The reptiles put to sea repeatedly, and for more than 100 million years their marine representatives—the icthyosaurs, plesiosaurs, and mosasaurs—ruled the waves; today, the largest marine animal (indeed, the largest animal of all time) is the blue whale, an air-breathing mammal like ourselves.

Yet side by side with the highly evolved invaders from the land, the undemanding sea continues to play host to thousands of species of "living fossils"—creatures changed little, if at all, from their ancestors of eons ago. A few parallels exist on land, of course: the opossum has not altered much in 100 million years, and the platypus, perhaps, for even longer. Yet compared to the sea's living fossils, they are newcomers. The coelacanth fish found a generation ago off East Africa differed little from skeletons chiseled out of rocks 250 million years old. In 1952 specimens of *Neopilina,* a very primitive mollusk, were fished up from the bottom of an 11,000-foot trench in the Pacific; they represented a group of animals supposedly extinct for 300 million years. And in 1965 David L. Pawson of the Smithsonian Institution, while dredging for marine organisms off Antarctica, found a kind of sea cucumber (a relative of the starfish) that not only resembled certain 300-million-year-old fossils but was apparently of the same species.

Moreover, the ocean's living fossils are not all obscure creatures. Multitudes of coral animals build reefs much as they did when the land was still a barren desert. The sea lamprey, a relic of the most primitive group of fish, made its way into the Great Lakes some years ago and multiplied so successfully that it almost exterminated several more advanced species of fish before it was brought under control. The horseshoe crab still litters beaches with its shells, as it has done for 400 million years.

In 1957 a Soviet oceanographic expedition discovered a clue to what may prove to be the most remarkable living fossil. With an underwater camera the Russians photographed the bottom of the equatorial Pacific some ten thousand feet down. One of the prints showed a series of trails, resembling four-inch-wide tire tracks, in the ocean-bottom ooze — trails that closely resembled fossil tracks discovered earlier in sandstone half a billion years old. It is just possible that both sets of tracks were made by trilobites, arthropods that between 500 and 600 million years ago were the most advanced animals on earth. Presumably the last trilobite died 300 million years ago — but I would not care to bet my life that one day a survivor will not be dredged up. The sea is very large and very deep, and there is still a lot we don't know about what lies within it.

Just as the sea preserves its many forms of life, it exerts a continual and powerful influence upon life on land through its role in producing weather, and the "averaged" weather patterns for particular regions, which we call climate.

Weather — and climate — happen basically because

some parts of the earth are much hotter than others. The difference in temperature between the equator and the poles sets in motion a gigantic "heat engine," comparable in its fundamental features to a steam turbine, whereby heat is transferred from the engine's "boiler" (the tropics) to its "condenser" (the polar regions). As in a turbine, the flow of heat from one place to another performs work—meaning in this case that it keeps both the atmosphere and the oceans in constant motion, in the process bringing different patterns of wind and rain, sunshine and snow, to different parts of the globe. And it is the ocean that, in one way or another, is responsible for a large proportion of the heat transfer that keeps the engine turning over.

A fair amount of heat is carried by warm currents such as the Gulf Stream and the Kuroshio of the Pacific (the name means "black current," from its deep blue color). These move north from the tropics, even as in the Southern Hemisphere the warm Brazil and Mozambique currents move south. The volume of the sea currents, while enormous by human standards, is puny in comparison with that of the great air currents (*e.g.*, the trade winds), which share the work of heat transfer. Yet the proportion of heat that the sea currents carry is much greater than their volume would suggest because the specific heat of water is far higher than that of air. For this reason, a bucketful of water can hold— and therefore transport—as much heat as a sizable roomful of air.

Even the fact that the atmosphere can carry as much heat as it does is largely due to the existence of the oceans. Much of its freight of energy is transported in the form of "latent heat," produced by the evaporation of sea water.

When water evaporates and becomes vapor, it absorbs heat (to verify this, sit in a draft with wet clothes on). Thus the water that evaporates from the oceans (at the rate of millions of tons an hour in their warmer regions) takes heat with it. This heat does not warm the atmosphere immediately. Instead, it remains locked up ("latent") until the water vapor condenses into rain or snow—frequently hundreds or thousands of miles north or south of the point at which it was formed.

The heat energy released by the condensation of water vapor is enormous, because water's "heat of vaporization" is quite as remarkable as its specific heat. The formation of a single cubic foot of rain water liberates enough heat to raise the temperature of 100,000 cubic feet of air by 15 degrees. The transfer of heat energy away from the tropics by evaporation and subsequent condensation of sea water is largely responsible for the fact that the stormiest regions of the earth— *i.e.*, those in which the atmosphere is most vigorously in motion—are not, as one might expect, along the equator, where the supply of heat is most copious, but well to the north and south, in the zones miscalled temperate.

The oceans' importance in providing the earth with a tolerable climate can be pointed up by imagining what things would be like if the sea did not exist. In the first place, the equatorial regions would be a great deal hotter, and the poles far colder, than they are now.

This accentuated contrast would force the atmosphere, carrying the entire heat load without benefit of ocean currents or latent heat, to move with inconceivable violence, raising winds that would make a hurricane seem like a spring zephyr. (Precisely such winds are believed to blow on hot, waterless Venus.) But these catastrophes would not trouble living organisms, of course; without the oceans, they would not exist.

Besides moderating equatorial and polar temperatures, powering storms in the temperate zones, and sharing the burden of heat transfer, the sea markedly affects the climates of particular regions. It does so by serving as a sort of "reservoir," which can absorb great quantities of heat in some regions and seasons, and release them at other times and places.

This capacity derives in part from the high specific heat of water, but there are other contributing factors. Because water is relatively transparent to the sun's radiation, the ocean can absorb heat at depths of up to several hundred feet. Moreover, wave action (and sometimes other mechanisms) stirs the ocean's upper layers, carrying warm water downward and bringing up cooler water to be warmed in its turn. On land, by contrast, solar heat penetrates only a dozen or so feet downward. (As spelunkers know, the temperature in deep caves varies only minutely from day to night and from summer to winter.) The "heat reservoir" on land is less than one-tenth as deep as that of the ocean. Moreover, volume for volume, the land can hold far less heat than the sea.

The effect of all this is that the ocean heats up and

cools off much more slowly than does the land, and its temperature range is much narrower than that of the land. As a result, the ocean exerts a powerful moderating effect on the climates of adjacent land areas. How much of an effect depends on which way the wind is blowing; clearly, areas in which the prevailing winds blow from land out to sea will be little affected by the ocean's moderating influence, compared with those in which the air flow is from sea to land.

Because of global wind patterns the contrast in climates is especially marked in the temperate zones, where the prevailing winds blow from west to east. Northwest Europe, for example, has a maritime climate, with cool summers and mild winters. In addition, its over-all climate is considerably warmer than one might expect, thanks to the Gulf Stream, whose waters carry a certain amount of tropical heat as far north as the sea between Norway and Greenland. In sharp contrast to this, the continental climate of eastern and central North America has hotter summers and colder winters than Europe does, and latitude for latitude, a distinctly colder climate over-all. Great Britain is on the same latitude as bleak Labrador, yet its climate (for all that tourists have said about it) is far more hospitable. And New York City, with its fairly extreme temperature changes, is on a latitude with balmy Naples.

A similar contrast is found on the shores of the Pacific, with modifications due to geography and topography. In the east the zone of maritime climate is much smaller than that of Europe. The warm Kuroshio, blocked by the Aleutian island chain, cannot penetrate as far north as the Gulf Stream does, so western Alaska is a great deal colder than coastal Norway. Farther south the ocean's moderating influence is checked by the Rockies and Coast ranges, so that only a narrow coastal strip of Oregon, Washington, British Columbia, and the Alaska panhandle has a maritime climate.

Along the Pacific's western shores, geography produces even more continental climates than it does in the comparable regions of North America, because the ocean's influence is less marked. To the west and northwest lies the enormous land mass of Eurasia; to the southwest the great Himalayan mountain system blocks off warm winds from the Indian Ocean. Thus Peiping, for example, has both hotter summers and colder winters than Boston, which lies in about the same latitude.

South of the European and North American zones of maritime climate the ocean's influence changes considerably. Here the warm currents have been dissipated, being replaced by the cold, southward-moving California and Canaries currents; water temperatures are further reduced by a rather complicated process called upwelling, in which surface waters move away from the shore to be replaced by colder waters rising from below. The result is a narrow zone of remarkably cold water along the coast. Thus, summer ocean temperatures off southern California are considerably lower than those off Cape Cod, which lies several hundred miles to the north—but on the opposite shore of North America.

This chilly strip of ocean considerably moderates summers along the Pacific Coast. In San Francisco July temperatures average 17 degrees cooler than those in Sacramento, only a few score miles inland. In addition, the cold coastal water chills warm air moving in from the west and condenses much of the moisture in it, giving San Francisco its famous fog—and contributing to Los Angeles' infamous smog. Comparable cold currents shape the climate in coastal Portugal and Morocco, in central Chile, and in a few other places.

The cold currents, together with more important atmospheric mechanisms, contribute to the hot, dry summers characteristic of these regions. Some hundreds of miles nearer the equator, the summer lasts year-round—which is to say that the land becomes a desert, its coastal sections often dampened by fog, but almost never visited by rain.

The most remarkable of these sometimes foggy deserts lies in northern Chile and southern Peru. It is one of the hottest places on earth and quite possibly the dryest: one settlement reported less than an inch of rain in forty years (New York City gets some forty inches of rain in one year). The cold current that helps create this forbidding region is the northward-flowing Humboldt, whose waters originate near Antarctica and whose influence, like that of the California Current, is strengthened by upwelling near the coast. Occasionally, however, the Humboldt, for unknown reasons, moves out to sea, to be replaced along the coast by a warmer current moving south from the equator, called El Niño.

The warmer waters set off torrential rains on land, which strip away the barren soil and reduce adobe buildings to heaps of mud. (El Niño's even more catastrophic effect on marine life will be considered in Chapter II).

While the ocean most markedly influences coastal climates in the areas mentioned, it plays at least a minor role on occasion almost everywhere that the sea is much cooler or warmer than the land. For example, the summer monsoon, which annually brings the rains to the Indian subcontinent, is essentially a scaled-up version of the sea breeze that cools a Cape Cod beach on a hot July afternoon.

Finally, the ocean is the only ultimate source of atmospheric moisture—and thus of every drop of rain or flake of snow that falls on land. Not every region adjoining the sea enjoys copious—or even skimpy— rainfall, but every region, without exception, that is isolated from the sea, whether by distance or (more often) by other causes, is desert or close to it. The Gobi Desert is about the same distance from the equator as the fruitful corn fields of Indiana, but while Indiana receives plentiful precipitation from moist winds moving up from the Gulf of Mexico, the Gobi is walled off from the Indian Ocean by the Himalayas. Similar deserts or semideserts, lying in the "rain shadow" of mountains that intercept oceanic moisture, are found in the Great Basin of the western United States, in much of southern Argentina, and in a broad belt across Central Asia, of which the Gobi is only the easternmost part.

Because the ocean is known to play so large a role in climate today, it is not surprising that many climatologists assign it a major role in climatic changes of the past. It is known, for example, that the earth's climate over the past two or three million years has been considerably colder and dryer than at most previous periods. Coal deposits in Antarctica testify to a rich vegetation where there is now barren rock and ice; fossils show that palm trees once flourished in Greenland. Periodically, however, the earth slowly cools to the point where great sheets of ice, like those that now blanket Antarctica and Greenland, cover large portions of the temperate-zone lands: this was true as recently as twelve thousand years ago.

The last such cooling-off period began perhaps forty million years ago, although the ice sheets did not appear until much later. And most geologists ascribe at least part of the cooling to changes in the extent of the sea.

Like every other physical feature of our planet, the oceans have not always been the same size or shape. Forty million years ago they covered a good deal more of the earth's surface than they now do. The major ocean basins were probably about the same size as they now are, but shallow seas covered the Mississippi Valley as far north as St. Louis; parts of North and West Africa were underwater, as was nearly all the Middle East; India was an island off the coast of Asia; and Europe was a group of islands west of Siberia.

All these shallow seas greatly expanded the upper layers of the ocean—and thereby enlarged the heat reservoir that those layers constitute. Thus, in nearly every part of the world the climate was under the ocean's moderating influence—the more so because there were almost no high mountain ranges to block ocean winds from the continental interiors. The polar regions had a rather mild, temperate climate while today's temperate regions were tropical or subtropical. With less contrast in temperature between the poles and the equator the atmospheric heat engine worked sluggishly; winds were gentle and storms were few.

A recession of the sea seems to have played a large part in changing this idyllic picture. The land gradually folded itself into the major mountain chains and the deep ocean trenches that we now know. Some geologists have also speculated recently that the mountain-building, in effect, squeezed the continents into a smaller area, thereby enlarging the area of the ocean basins. Whatever the precise cause, the shallow seas gradually drained off the land and into the deeper ocean basins, creating colder, more continental climates and setting the stage for the ice ages that followed.

Along with its role in bringing about long-term climatic changes on land, the sea also seems to influence short-term fluctuations, although the mechanisms by which it does so are still obscure and controversial. There seems no doubt that year-to-year variations in the temperature and volume of the Gulf Stream, as well as in the cold currents that flow into the Atlantic from the north, have a considerable effect on the weather—particularly by influencing the movements of storms over eastern North America and western Europe. But this, in turn, raises the question of what causes the changes in the ocean currents, and the answer seems to be — changes in the atmosphere. When it comes to weather, there is a continuous and complex interaction between the oceans and the atmosphere, and we are a long way from being able to say which is cause and which effect — if, indeed, the terms have any meaning here.

If the sea, through its influence on weather and climate, exerts a major and continuous influence on man's terrestrial habitat, it also affects us more subtly through its importance as a food resource. Man's harvest of fish alone, discounting shellfish and such exotic marine delicacies as sea cucumbers, exceeds sixty million tons a year, and as we shall see later, it could perhaps be increased to several times that figure. In bulk, sixty million tons of fish is still an inconsiderable part of the world's food supply, but it is a crucial one, since the primary nutritional element in fish is protein. And protein deficiency is the chief nutritional illness affecting mankind; the debilitating (and sometimes fatal) disease kwashiorkor — caused by inadequate protein in the diet — is thought to affect at least 100 million people in underdeveloped tropical countries.

Curiously, relatively few other land animals have tapped the sea as a source of food. To be sure, the whales and seals proliferated (until decimated by man) on a diet of sea life — but whales are wholly marine animals, and seals hardly less so, as anybody will appreciate who has ever watched a seal maneuvering on land.

The outstanding exception is the birds, enormous numbers of whom find their living at sea, and do so, for the most part, without abandoning any of their essential "birdishness." Of the seven-thousand-odd species of birds, only some five hundred are dependent on the sea for their food, and these include the various species of ducks that feed only in bays and salt marshes. But in numbers the sea birds strain the imagination. Scientists studying bird populations off the west coast of South America have estimated that one flock of cormorants numbered something like a million — and that is one flock of one species in one region, albeit a region peculiarly rich in sea birds. When we add up all the flocks of cormorants, gulls, terns, auks, pelicans, and penguins, the tiny sandpipers that scuttle back and forth in front of the curling surf line on the beaches, the petrels and albatrosses that spend almost their entire lives at sea, we are clearly talking about a very substantial fraction of the world's bird population. The birds' harvest of fish is believed to equal that of all man's commercial fisheries put together.

The sea is also an immense storehouse of minerals (one is tempted to use the old cliché "inexhaustible reservoir" — but bitter experience with "inexhaustible" natural resources on land has taught us caution). Sea water, evaporated in shallow pools, has always supplied a sizable proportion of man's salt; today, the sea is a major source of the light metal magnesium — essential for airplane construction — and of bromine, used among other things for making antiknock gasoline. Millions of dollars worth of diamonds are being culled from sea-bottom sand and gravel off the coast of south-west Africa, and immense submarine deposits of phosphate and manganese nodules await only the development of economic methods for "mining" them.

In a few spots, such as the coast of Brittany, the tides (which there rise and fall up to thirty-five feet) are being harnessed as a source of hydroelectric power. And if the physicists ever work out a way of controlling the nuclear-fusion processes of the hydrogen bomb, there is enough heavy hydrogen in the oceans to fulfill our need for energy for tens of thousands of years.

Yet above and beyond its material riches, its intricate climatic influences, and the fascinating complexities of its own physical and biological processes, the sea is still somewhat akin in our minds to the dark, forbidding Ocean River of the ancient geographers. For all the hundreds of millions of dollars spent on oceanic exploration over the past century, its depths remain by all odds the largest terra incognita on our planet. And not always the depths. Only a few years ago a U.S. Navy vessel on a routine trip from Panama to Miami took soundings on a hitherto unknown mountain under the Caribbean that reached to within sixty feet of the surface. Not so many years before that, an oceanographic expedition discovered, a hundred yards or so below the surface of the Equatorial Pacific, a major new current flowing directly counter to the surface currents. It is anyone's guess how many spectacular topographic features, how many strange currents and tides, how many curious or prehistoric forms of life, still lie unnoticed beneath the ocean waves.

Moreover, for all of us who explore it in boats or books, sail it in sloops or ocean liners, or merely watch it from the shore, the sea is a major untamed—and untamable—force of nature. Despite the ingenuity of modern shipwrights and the skill of modern navigators, with their electronic loran, radar, and sonar, the sea's wind-whipped waves each year claim a score of ships that prove unequal to its power. Each winter its pounding breakers gnaw away at the land, stealing a few inches a century from the rocky headlands of Maine, as much as three feet a year from the sandy bluffs of Cape Cod, but always pressing inward on the continents, and ultimately destined to engulf great regions that are now dry land, even as it did forty million years ago. Our most powerful meteorological disturbances, the hurricanes—each one liberating energy equivalent to a dozen or more of our puny H-bombs—are born at sea and dissipate rapidly when they move over land, leaving a trail of smashed breakwaters, snapped telephone poles, and roofless houses. Earthquakes and volcanic eruptions are fearsome enough on land, but at sea their destruction is multiplied by the mighty tsunamis (miscalled "tidal waves") that they stir up.

Thus the sea, even as it beckons and fascinates us, also awes and terrifies us. Its encircling embrace, so often benign and sustaining, can become a kiss of death. Inescapable, irresistible, and eternal, it reminds us, should that be necessary, of our own limitations, of the fact that for all our civilization, science, and technological gadgetry, we are still rather small and feeble animals in a large and by no means friendly world.

CHAPTER II
AT HOME
IN THE SEA

Along the coast of the Bahamas and southern Florida, as in many other coastal areas of the Caribbean and Gulf of Mexico, lie beds of turtle grass, which, in the words of the naturalist William M. Stephens, "spread over the shallow bottom like a mottled green carpet." This remarkable plant is not a seaweed but a true grass whose ancestors, evolving on land some sixty million years ago, subsequently returned to the sea—one of the very few advanced plants to do so.

It is easy to inspect turtle-grass flats from a rowboat, for they flourish chiefly in protected bays and lagoons, in the shallows below low-tide mark. Seen through the clear, warm water, a turtle-grass flat seems almost devoid of animal life; only a scattering of sea cucumbers, starfish, and sea urchins, a few fish, and perhaps a conch break the monotony of the undulating grass blades. But appearances are deceptive. Although living turtle grass is eaten by only a few animals (its chief former consumers, the sea turtle and the manatee, have been almost wiped out by man), it is covered with nourishing bacteria, protozoans, and one-celled plants (algae). These, together with the debris from broken and dying blades, support a rich population in the grass and in the mud beneath it.

Recently Donald R. Moore of the Institute of Marine Sciences in Miami sampled the life in turtle-grass banks at several locations in Biscayne Bay. Using fine nets and a "plug sampler," which can lift up a section of bottom a foot or more deep, he found an average of some 72 small shrimp per square yard—which adds up to well over 200 million per square mile. The figures for mollusks were even more startling: some 8,000 small clams and 200,000 tiny snails per square yard, or nearly 100 *billion* mollusks per square mile!

Other important inhabitants of the turtle-grass flats are the young of many larger fishes and crustaceans, some of them commercially important—sea trout, pompano, barracuda, mullet, stone crab, and spiny lobster. They take advantage both of the rich food supply and of the cover that the grass provides against larger predators.

Perhaps the most singular member of the turtle-grass community is the small, sinuous pearlfish, which lives within the body cavity of a sea cucumber. It usually enters its home by swimming backward into the cucumber's anus, and it supports itself by browsing on the cucumber's internal organs. This does not appear to incommode the cucumber; indeed, many of these animals will eject their viscera to distract a predator, and after escaping, will grow a new set of organs at leisure.

Turtle-grass banks benefit from a favorable combination of the physical factors that govern life in the sea and elsewhere: light, temperature, and chemical raw materials—the carbon dioxide and mineral nutrients (chiefly nitrates and phosphates) required by plants, and the oxygen needed by both plants and animals. Differences in the availability of these essentials determine which organisms, and how many of them, can live in a particular patch of ocean.

The main factor in the productivity of the turtle-grass banks is an ample supply of mineral nutrients. This is due partly to runoff from the adjacent land (sewage is an abundant source) and also to the shallowness of the water. In the open ocean the supply of minerals near the surface is usually depleted by incorporation into the bodies of plants and animals, which, as they die, carry the nutrients downward to the lightless regions where plants cannot grow. In shallow water, however, the bottom is still within reach of the sunlight, enabling plants to re-use the mineral "fallout" in an almost endless cycle.

Even more productive than the turtle-grass communities are the chilly waters off southern Peru. As already noted, the coastal lands there are desert—but not so the sea. Upwelling waters restore to the lighted surface layers the nutrients that would otherwise accumulate in the depths, at the same time keeping the surface waters cold. Other things being equal (they seldom are), cold waters are more productive than warm. Myriads of microscopic plants and animals nourish shoals of anchovetas; bonitos pursue the anchovetas, and tuna and sea lions eat the bonitos. Clouds of cormorants, pelicans, and gannets swoop down on the anchovetas from above, sometimes whipping the water into foam as they gather an estimated four million tons of fish a year. Human fishermen are there as well, hauling in more than ten million tons annually; other humans collect guano, a high-grade fertilizer that the birds deposit by the thousands of bushels on the arid islands where they nest.

In sharp contrast to these teeming communities, the Sargasso Sea in the central North Atlantic is both warm and poor in minerals. It is remote from any rejuvenating runoff from the land, and its surface layers receive no nutrients from the depths. For reasons having to do with the over-all flow of oceanic currents the Sargasso waters undergo downwelling rather than upwelling. The deep, clear indigo of the Sargasso testifies to a sparse population of microscopic organisms—and necessarily, very limited numbers of the larger animals that feed on them. "Volume for volume," says the oceanographer John H. Ryther, "the Sargasso Sea is the clearest, purest and biologically poorest ocean water ever studied."

Yet even here there is life. Standing in the bow of a West-Indies-bound freighter, one can watch the flying fish spurt out of the sea just ahead of the cutwater and glide for twenty or thirty feet on their winglike fins before dropping back into the water. Long windrows of sargassum weed stretch to the horizon (the meager factual foundation for the old sailors' tales of a sea of weed in which trapped ships drifted forever).

Unlike most seaweeds, which attach themselves to rocks or to the sea bottom along the coast, the sargassum apparently floats freely, buoyed up by its bean-sized air bladders. Moreover, it has evidently existed this way for millions of years—long enough for a specialized population of animals to evolve in and upon it. Sea slugs (shell-less snails) and tiny crabs crawl about these miniature rafts, and the sargassum fish swims amid its tangled shoots—all camouflaged by their yellowish-brown color and by grotesque flaps of tissue that mimic the ragged patterns of the weed itself.

Though the Sargasso Sea is a marine desert compared with the rich waters off Peru, it is still fertile enough to serve as a spawning ground and nursery for the entire eel population of western Europe and eastern North America. In the autumn of every year millions of eels make their way down-river and out into the Atlantic, where by routes and navigational mechanisms still unknown they travel to the Sargasso to spawn and die. Their larva, leaf-shaped and quite uneellike, make their way back again, feeding, growing, and metamorphosing as they go, until at last, one to three years later, the young eels swim up-river to begin adult life.

All these marine habitats, and others equally well known—the coral reef, with its strangely shaped masses of limestone and its brightly colored fish; the between-tides zone of rocky coastlines, streaming with seaweeds and dotted with mussels, limpets, and periwinkles—are in the ocean's upper layers, which are most accessible to man. But most of the sea is lower down, where its life cannot be studied, or even viewed, without highly specialized equipment.

Here, in perpetual darkness, swim fish with luminous "portholes" along their sides; crimson shrimp evade their enemies by emitting "smoke screens" of glowing fluid. One oceanographer has compared these displays to a Fourth of July celebration. Nor are simpler forms of life lacking: starfish and worms, many of them differing little from shallow-water species, crawl along the bottom and burrow in its mud. The oceans' dark depths are thinly populated, but they are so immense in bulk—something like 90 per cent of the oceans' volume—that they may well contain as many living organisms as the more hospitable, but far thinner, upper layers.

Various factors shape life in these marine communities, as well as in the many others we have not mentioned (the deep waters, for example, are divided into the moderately deep bathyal zone and the deeper abyssal zone; some oceanographers recognize an even deeper hadal zone).

All living organisms require a source of energy, and in the sea, as on land, the basic source of energy is the sun's radiation. The chemical catalyst chlorophyll allows plants to transform this solar energy, through photosynthesis, into the chemical energy of starches, sugars, fats, and proteins, which they manufacture chiefly from water and carbon dioxide, plus small quantities of other nutrients. This chemical energy, in turn, supports the animal kingdom, which exists by consuming plants or plant-eaters.

Apart from a few places, such as caves, almost all sections of the land are exposed to at least some light, and the energy is available to all creatures dwelling on it. In the sea, by contrast, photosynthesis can occur only in the relatively shallow surface layer, while marine life exists at all ocean levels. The degree to which sunlight can penetrate sea water depends upon the angle of the sun, the amount of suspended matter, living or dead, and many other factors. Significantly, the ocean is, in general, most "transparent" to the sun's blue-green wave lengths, which are precisely those that chlorophyll absorbs most effectively. But even under optimum conditions—clear water, the sun directly overhead, and so on—the sea's lighted zone is narrow. When the Norwegian oceanographer B. Helland-Hansen exposed photographic plates at a depth of some 250 fathoms off the Azores, they were only slightly blackened even after forty minutes. Broadly speaking, we can say that nearly all the ocean's primary production of chemical energy occurs in its uppermost one hundred yards. In the immense regions below, animal life (there are no living plants) is entirely dependent on such food as drifts down to it.

If we visualize marine animals as catching their food "on the fly," it seems a rather chancy way of finding a living. The picture is misleading, however. The great majority of food organisms are very small, and therefore, relative to their weight, they have an enormous surface area to slow their fall. It has been estimated that a copepod—a small crustacean—would fall at the rate of only two feet a minute, or a mile every couple of days. And a copepod, as marine organisms go, is rather large (meaning that it is visible to the naked eye), so there can be no doubt that the smallest plants and animals would require weeks to reach the bottom—allowing plenty of time for their consumption on the way down.

This "thinning out" process necessarily reduces the available food at each descending level. Immense areas of the ocean bottom are, indeed, covered tens and hundreds of feet deep with mud composed of the minute "skeletons" of marine creatures that have fallen from above—but these deposits represent tens and hundreds of thousands of years of "fallout." As a general rule—there are a number of exceptions—both the quantity and diversity of marine life decreases sharply with depth.

Temperature is considerably less limiting than light in its effects on marine life. Unlike the land, there are no places in the sea either too cold or too hot to support life. The frigid waters off Antarctica probably nourish more life per cubic yard than any other marine region; yet, on the adjacent land, plants are almost unknown, and such animals as do exist are almost all totally dependent on the sea for food.

Yet differences in sea temperature do exert very pronounced effects on the particular species inhabiting a particular region of the ocean. Most species of fish are physiologically adjusted to particular temperature limits and will die if they find themselves in significantly colder or warmer water. The tissues of the cod, a cold-water species, will actually coagulate ("cook") if exposed to water as warm as the Gulf Stream off Florida.

Variations in temperature are significant only in the ocean's uppermost layers; below a few hundred fathoms the water is uniformly cold—at best, a few degrees above freezing. During the cruise of H.M.S. *Challenger* (1872–76), a landmark in oceanographic research, the ship's officers used bottom ooze dredged up from beneath the tropical waters to "ice" their champagne!

Near the surface, sea temperatures range from 90 degrees or more in parts of the Persian Gulf to below 32 degrees near the poles (sea water, because of its salt content, does not freeze until it is several degrees "below freezing"). As on land, surface temperatures vary with changes in the amount of solar heating due to latitude and to the seasons. But, also as on land, large areas of the ocean are either warmer or colder than they "should" be, taking into consideration solar radiation alone.

For one thing, the relatively slow heating of water ensures that maximum and minimum sea temperatures will lag considerably behind maximum and minimum radiation. In land areas remote from the sea January is normally the coldest month and July the warmest (this is in the Northern Hemisphere; in the Southern, of course, the reverse is true). But at sea—and in regions of maritime climate—surface temperatures almost invariably hit bottom in February and peak in August.

The most dramatic peculiarities in ocean temperatures, however, occur in the great warm and cold currents. We owe much of our knowledge of these currents to one of the first, and greatest, oceanographers, Lt. Matthew Fontaine Maury, U.S.N., a man equally brilliant at planning research, interpreting its results, and presenting these results in a lucid and forceful style.

For sheer clarity and eloquence the first paragraph of his book *The Physical Geography of the Sea* has never been surpassed—and seldom equaled—in scientific writing:

There is a river in the ocean. In the severest droughts it never fails, and in the mightiest floods it never overflows. Its banks and its bottoms are of cold water, while its current is of warm. The Gulf of Mexico is its fountain, and its mouth is in the Arctic Sea. It is the Gulf Stream. There is in the world no other such majestic flow of waters. Its current is more rapid than the Mississippi or the Amazon, and its volume more than a thousand times greater.

Maury's description of the Gulf Stream, and of the ocean's other great rivers, was the product of what was probably the first large-scale co-ordinated research project in history. Before his time, knowledge of winds and currents was for the most part a trade secret of ship-masters and mariners. Maury began by collating as many of these observations as he could get hold of, "putting down on a chart the tracks of many vessels on the voyage, but at different times. . . . [and] projecting along each track the winds and currents encountered."

This first chart, he recognized, was "meagre and unsatisfactory." He thereupon entered into correspondence with American sea captains; "their attention was called to the blank spaces, and the importance of more and better observations . . . was urged upon them.

"They were told that if each one would agree to cooperate in a general plan of observations at sea, and would send regularly, at the end of every cruise, an abstract log of their voyage to the National Observatory at Washington, he should, for so doing, be furnished, free of cost, with a copy of the charts and sailing directions that might be founded upon those observations . . .

"The quick, practical mind of the American shipmaster took hold of the proposition at once . . . So in a little while, there were more than a thousand navigators engaged day and night, and in all parts of the ocean, in making and recording observations according to a uniform plan . . ."

The picture of the ocean's surface currents that Maury pieced together has been much refined since his day, but not fundamentally revised. He plotted a series of enormous whirls in the oceans, turning clockwise in the Northern Hemisphere and counterclockwise in the Southern. Thus, in the North Atlantic, where for a number of reasons the pattern is clearest, the warm North Equatorial Current flows west somewhat north of the equator. Turning northwest, it becomes the Gulf Stream, which flows successively north and then northeast. Part of its flow pushes as far north as the Arctic Ocean; the remainder, turning east and then southeast, brings its benign influence to the coasts of England and France.

Chilled in these northerly regions (and with the addition of colder water from below), the current moves south as the cool Portugal and Canaries currents until at last it turns westward along the equator to complete the circuit.

The pattern in the South Atlantic is less pronounced, but otherwise almost a mirror image: the warm Brazil Current flows south from the equator on the west and the cold Benguela Current pushes north past southwest Africa on the east. The same patterns are repeated in the Pacific. In the Indian Ocean, whose basin straddles the equator, the currents are rather more complicated. There, major seasonal shifts in the prevailing winds cause some currents actually to reverse direction between summer and winter.

Understanding the patterns of warm and cold ocean currents is important not only to the navigator but also to the fisherman. Water temperature determines which species are to be found in a particular area; it also appears that cold waters are usually richer in life than warm.

One reason is that cold water can contain more of the essential raw materials, oxygen and carbon dioxide, than warm water can. (To verify this, let a bottle of soda pop warm in the sun before opening it.) Another reason is that low temperatures, though they slow the chemical processes of growth, apparently slow those involved in aging much more markedly. Thus cold-water organisms take longer to mature — but a great deal longer to die. In the nearly freezing waters off Antarctica, says the naturalist Robert Cushman Murphy, "many more successive generations of each species of marine organism live contemporaneously than exist in warmer waters."

On the other hand, life in warm waters is considerably more diverse (measured by the number of different species) than is that in cold waters. This is sometimes explained by the fact that warm-water organisms, by maturing faster, run through more generations in a given period of time and thus have evolved faster and more diversely. I myself do not find this altogether convincing, for if evolution—the process by which organisms become more perfectly adapted to their environment—proceeds more rapidly in warm waters, the extinction of the poorly adapted species should proceed no less rapidly. I suspect that at least part of the true explanation lies in the fact that the really cold regions of the ocean, as of the land, are of relatively recent vintage.

During almost all the earth's history tropical or near-tropical conditions have extended nearly to the poles on both land and sea. The great icecaps of Greenland and Antarctica, which do so much to chill the polar regions, were formed not much more than five million years ago—only yesterday on the evolutionary time scale. Thus it may well be that relatively few species of organisms inhabit cold waters because too little time has elapsed for widespread evolutionary adaptation to that habitat.

Whatever the reasons for differences between cold-water and warm-water populations, it is a fact that sudden changes in water temperature can wreak havoc among marine organisms. We have already noted the drastic climatic shifts on the Peruvian coast when the chill waters of the Humboldt Current are replaced by the balmy flow of El Niño. At sea the watery "heat wave" kills fish and other marine organisms by the billions; the hydrogen sulfide released by their rotting bodies can blacken the white-lead paint of ships, giving this stretch of ocean the seaman's name of "Callao Painter," after the Peruvian port of Callao. The millions of sea birds, deprived of their normal food, die or seek richer waters; their production of guano, in normal years approximately 100,000 tons, drops to almost nothing.

Less often, vagrant cold currents can be equally destructive. In 1882 a flow of Arctic water is believed to have intruded into the ocean off the northeastern United States. The result was millions of dead tilefish (then an important commercial species); one fishing schooner reported that it had sailed "for about 150 miles through waters dotted as far as the eye could reach with dying fishes."

Even under the most favorable conditions of light and temperature, marine organisms cannot grow without an adequate supply of raw materials. Carbon dioxide (CO_2) and oxygen dissolve into the sea from the atmosphere. The CO_2 is used only by plants; indeed, it is produced by living animals — and by the decay of dead plants and animals. It is "needed" only in the ocean's upper, sunlit layers where photosynthesis can take place, which is to say, precisely those layers that are closest to the atmosphere.

Thus the supply of CO_2 to marine plants poses no scientific problem. Oxygen, which is needed wherever animal life exists, is a different matter. And since animal life unquestionably exists even in the deepest part of the ocean, it is clear that oxygen must be present there, too. How does it get from the surface to the depths? The ocean is constantly being "stirred" by the winds and surface currents, but these processes affect only the uppermost layers. Evidently, then, there must be a subsurface circulation that carries oxygen-rich water down to the depths.

Oceanographers have found immense currents of precisely this sort. Unlike the major surface currents, their flow is far too sluggish to be measured directly by even the most sensitive modern meters; while the Gulf Stream travels a maximum of seven miles an hour, deepwater currents do well to cover that distance in a day. Nonetheless, by precisely measuring small differences in the temperature and salt content of water at various depths oceanographers have been able to identify various "water masses" and map their leisurely flow beneath the surface of the sea. These same variations in temperature and salinity produce differences in density, which in turn provide much of the motive power to keep the currents flowing.

The patterns of the deep currents are too complex for any summary description. This is because temperature and salinity tend to produce opposite effects, often canceling each other out. Surface water is less dense in warm parts of the ocean (water, like most substances, expands with heating). At the same time, the high rate of evaporation makes the surface waters saltier (therefore denser) than deeper waters. This is especially true in regions, such as the Sargasso Sea, that are remote from the fresh water supplied by rivers. Near the poles, on the other hand, surface water is colder (and therefore denser), but also less salty (and therefore less dense), due both to a lower rate of evaporation and to the addition of fresher water supplied by melting ice.

This seeming standoff is overcome in two ways. First,

warm, salty water moving toward the poles loses much of its heat on the way, gaining in density as it does so. Second, part of the cold, polar water will "lose" some of its fresh water through the formation of ice. Crystals of sea ice, though they retain a certain amount of brine on their surfaces, are themselves almost pure fresh water, so the water "left behind" when the sea's surface freezes will be considerably saltier than it was before.

Evidently, then, the water that is dense enough to "fall" from the surface into the depths will be chiefly of two kinds: warm water of high salinity that has been chilled by moving into colder regions, and cold water of relatively low salinity that has gained in salt content through ice formation. In either case it is obvious that the major transport of water from the surface to the depths must take place in and around the polar regions. Providentially, it is precisely in these regions that the surface waters are richest in oxygen. The Antarctic waters, indeed, contain so much oxygen that they can support one of the most remarkable group of animals ever discovered. The icefish, as they are called, are almost devoid of hemoglobin, the substance that makes blood red (theirs is colorless) and that, through its capacity to combine with oxygen, permits the blood of other higher animals to transport relatively large quantities of oxygen. The icefish, however, seem to get along well enough on such oxygen as their hemoglobinless blood can carry—due apparently to the high oxygen content of their home waters. And these waters can be traced, at great depths, well north of the equator, supplying oxygen to deepwater organisms all the way.

In a few places, however, the ocean's "oxygen pump" breaks down. The outstanding example is the Black Sea. Because it receives the flow of so many rivers (including two of Europe's biggest, the Danube and the Don), its surface water is only about half as salty as that of other oceans. This layer "floats" on the saltier, denser, deep water, hardly mixing with it. Nor can the deep water receive an extensive supply of oxygen from the deep waters of the Mediterranean, since the Bosporus, which connects the two seas, is less than a hundred feet deep.

The lower layers of the Black Sea are so poor in oxygen that living debris drifting down from above cannot decompose completely. The decomposition that does occur is chiefly the work of bacteria that can exist without oxygen—many of them producing the poisonous gas hydrogen sulfide. This, and the lack of oxygen, prevents any other organism from living in the Black Sea at depths greater than five hundred feet. The sea's bottom mud is rich in partially decomposed organic matter; accumulations of such material over millions of years, it is thought, may be one of the ways in which petroleum is formed.

The mineral raw materials needed by ocean plants— mainly nitrates and phosphates—come from two sources. One, as we have seen, is runoff from the land into rivers, and ultimately into the ocean. It is partly for this reason that coastal waters are considerably more productive than most regions of the open sea.

The second source of minerals is the ocean depths

themselves. The rain of dead organisms and animal wastes depletes the upper layers of nutrients, so in the absence of counteracting processes (runoff from the land, or upwelling) the mineral content of the upper layers is held near zero. But it increases with depth to a maximum of around five hundred fathoms, below which it remains nearly constant all the way to the bottom.

Upwelling restores these mineral-rich waters to the surface zone where plants can utilize them. In part, upwelling is simply the consequence of the downward transport of surface waters in polar regions. For if millions of tons of water are moving down in one place, an equal quantity must be moving up somewhere else. At times "somewhere else" is far away, but at other times it is very close to the region of downwelling—as is the case in the productive seas off Antarctica, where the immense downward drift of surface water forces quantities of nutrient-rich deep water to the surface.

Downwelling occurs chiefly—though by no means exclusively—near the poles. Upwelling, on the other hand, takes place in many temperate or tropical regions, in particular, wherever ocean currents move away from shore. Off California, for example, the coast trends toward the southeast, while the California Current, because of the prevailing winds, moves toward the south, carrying surface waters away from the coast. This water must be replaced, since water seeks its own level, and the only source of replacement is the ocean's deeper layers. The same circumstances produce upwelling inshore of the Humboldt Current; in fact, upwelling is a major factor in the perennial chill of the waters off western South America. It has been determined that the Humboldt's temperature rises hardly at all in its long journey up the coast from Cape Horn to tropical Peru — which could not happen unless it was constantly receiving cold reinforcements from below.

A sort of upwelling occurs regularly during the winter in most temperate and cold regions, producing an annual cycle of ocean productivity. During the summer months the upper layers, warmed by the sun, grow less dense and "float" on the lower waters. Their nutrients are rapidly exhausted, so plant growth drops off sharply. During the winter the surface waters become colder and denser than the layers immediately below and "fall" downward, forcing the richer, lower water upward. At that season, however, the ocean plants cannot take much advantage of the improved nutrient supply. The days are short, and the sun is so low that much of its light is reflected off the water instead of being absorbed; moreover, the cold temperatures slow the metabolic processes of the plants. But in spring the longer hours of sunlight warm the surface waters and produce an almost explosive multiplication of marine plant life; enormous "blooms" of these minute organisms can turn immense sections of the ocean green or yellow or even red. Similar blooms also occur in areas (such as parts of the Florida coast) where large quantities of nutrients may be washed from the land into the sea by heavy rains. Upwelling in the Red Sea may cause the blooms of red algae that give it its name.

It is worth noting, finally, that upwelling is suspected of being another process responsible for the formation of petroleum. According to this theory, in zones of intense upwelling the fall-out of organic matter onto the ocean floor may be so profuse as to exhaust the oxygen content of the bottom waters, so that decomposition is sluggish, as it is, for different reasons, in the Black Sea. Alternatively, the rich organic layers may be buried and isolated from oxygen by turbidity currents — which are, in effect, undersea mud slides or sand slides. Certainly all, or nearly all, present oil deposits seem to occur in rocks that were formed under the sea; presumably, these mark prehistoric zones of upwelling.

Understanding the factors influencing ocean productivity is obviously of immense importance to the development and management of commercial fisheries. From this standpoint the chief limiting factor seems to be the supply of certain mineral nutrients. Carbon dioxide seems never to be in short supply in the lighted layers where plants can use it, and the same appears to be true of oxygen — apart from a few special cases already mentioned. But there is still a good deal we do not know about the role of other substances in the growth and reproduction of marine life. For example, the microbiologists Seymour H. Hutner and John J. A. McLaughlin have made extensive studies of the nutritional requirements of dinoflagellates, one-celled organisms resembling both plants and animals, which are an important element in ocean productivity. In sea water these organisms usually grow well enough, but Hutner and McLaughlin found it extraordinarily difficult to concoct a wholly artificial medium for their dinoflagellate culture. "Differences as small as one part per million in the concentrations of trace elements such as iron, zinc, magnesium, cobalt or copper can determine whether a particular species will reproduce or not," they reported. In particular, they found that dinoflagellates depended heavily upon a supply of vitamin B_{12}, which is used to treat pernicious anemia in human beings. The B_{12} molecule contains cobalt, suggesting that a shortage of this element might hold down productivity in regions where it would otherwise be high. "We know there are 'cobalt deserts' on land where livestock die for lack of this element in the soil," say the two researchers. "Perhaps some deserts in the sea have the same deficiency."

A number of physical factors also have an important impact on marine life along the coastal bays, estuaries, salt marshes, beaches, and rocky headlands. In estuaries and some bays, for example, the salty ocean water at the mouth mingling with the fresh water upstream sets boundaries on the living space available to marine organisms, most of which cannot tolerate marked changes in the salt content of their habitat.* Some, however, seem almost indifferent to salinity: the blue crab, well known to gourmets, thrives in water almost as salty as the open ocean or almost as fresh as a mountain stream. Even more remarkable are fish like the eel, salmon, shad, and striped bass, which spend

---

* The Baltic Sea, incidentally, is a sort of king-sized estuary. In its extreme eastern sections the water is fresh enough to drink, while in the narrow Kattegat, the Baltic's only outlet to the North Sea, the water is almost as salty as sea water.

part of their lives in salt water and the remainder in fresh.

Organisms that live in the zone between high and low tides are confronted by other problems. All of them must be able to cope with the periodic retreat of the waters upon which they depend for the necessities of life. Those that live on exposed coasts must also withstand the battering of the waves.

Many parts of the seas' borderland, such as the turtle-grass flats described earlier, are of special importance to marine life—and therefore to man. They are usually highly productive, and their products often include such commercially valuable species as oysters, clams, and crabs. Even more important, they serve as nurseries for the young of many commercial species of fish. Not only do they supply a rich diet for the young animals, they also furnish protection. In the open ocean there is no hiding place—no equivalent of the underbrush or grass in which small animals can escape their predators. Amid the tangled strands of a turtle-grass or eelgrass flat, or the winding channels of a marsh, however, the young fish has at least a fighting chance of surviving long enough to reach adult size—thereby acquiring some capacity to cope with the perils of the open sea. (Evidently, the bigger an organism is, the fewer are the predators that can feed on it; the most ferocious 30-pound barracuda would have trouble making a meal of a 500-pound tuna.)

Unfortunately, it is precisely the oceanic borderlands that have suffered most from man's disregard for his environment. Pollution and dredging have destroyed miles of oyster beds, converting that delectable mollusk from a plentiful, cheap food ("not worth an oyster," said Chaucer) to a luxury. Pollution has driven the succulent lobster from almost the entire length of Long Island Sound and made many clam beds sources of disease rather than of comestibles.

Even more severely threatened, perhaps, are the "wetlands"—the tidal flats and salt or brackish marshes that nourish and shelter so much young marine life and exert a positively hypnotic attraction on developers, land speculators, and short-sighted townships seeking more taxable real estate. Since they are "worthless" lands—except to fish, fishermen, and fish-consumers—they can be snapped up at bargain prices. Filled in with sand, mud, or garbage, the wetlands are transmuted into waterfront property selling for tens of thousands of dollars an acre.

A sizable part of San Francisco Bay has already been destroyed in this manner, enriching developers even as it impoverished marine life. And the largest wetland area in the United States, the Florida Everglades, already endangered by man's interference with its water supply, now faces the additional threat of a major airport. Construction of this facility, plus the commercial development that could be expected to go along with it, would enrich a few Floridians, destroy part of the Everglades outright, and pollute much of the remainder with sewage and jet exhaust. The effects on the marine life of the region could be catastrophic.*

---

* At this writing, the Everglades' jetport has been blocked for the time being, thanks to the efforts of conservationists in Florida and elsewhere. How long it will stay blocked remains to be seen.

The sea is so immense that we assume nothing man can ever do will seriously damage or even alter it. So far as the open ocean goes, there is some truth to this — though less and less with every year. But in the borderlands, so crucial to marine life, the shortsighted pursuit of the fast buck is destroying an irreplaceable natural resource.

CHAPTER III
GETTING A LIVING

A marine biologist once calculated that if every egg laid by every codfish hatched and grew to maturity, the Atlantic would become a solid mass of cod within six years. This mathematical exercise well illustrates two basic facts about marine life: its extraordinary fecundity and the equally extraordinary odds against its survival. In the sea, as on land, the basic patterns of life revolve largely around who eats and who is eaten—and in the sea it has been estimated that the chances of any individual organism living out its life without ending up inside some other organism are approximately one in ten million.

The fact that marine species can survive and flourish despite these odds suggests that most of them must be extraordinarily numerous and therefore, necessarily, extraordinarily small. This is quite correct. And the smallest and most numerous marine organisms are the plants, whose photosynthetic activities support all other marine creatures, from microscopic protozoa to 100-foot whales.

When most of us think of marine plants, we are likely to visualize seaweeds fringing a rock, or perhaps the windrows of eelgrass or turtle grass washed up on the beach after a blow. But nearly all these visible multi-celled plants grow on the ocean's fringes—and even there, account for only a small fraction of the sea's "primary production" of chemical energy through photosynthesis. One of the basic differences between marine and terrestrial life is that in the sea most of the vegetation is microscopic.

Like many other important scientific truths, this one emerged slowly. Though an ancient proverb had noted that the big fishes eat the little ones, it was a long time before it occurred to anyone to wonder what the little fish ate.

One of the early wonderers was Charles Darwin. During the historic voyage of H.M.S. *Beagle* the young naturalist made many of the observations about living creatures that he later wove into the theory of evolution. Trailing cloth nets astern at various times during the voyage, he collected "many curious animals," including numbers of copepods (tiny crustaceans now thought to be the most numerous many-celled animals in the world). He understood that such tiny creatures must provide food for such larger animals as the flying fish and "their devourers the bonitos and albicores [sic]."

Darwin realized that the copepods and other lower animals must in turn feed on the Infusoria, which we would now call protozoa. "But on what, in the clear blue water, do these Infusoria subsist?" he asked. His nets, made of loosely woven bunting, were not fine enough to answer the question.

Some ten years after Darwin had returned to England, the great German biologist Johannes Müller provided the beginning of an answer. Müller had apparently never heard of Darwin's work, nor, for that matter, was he particularly interested in marine plants. He was looking for the larvae of starfish, and to catch them he began dipping up sea water and pouring it through a sieve of cloth much more finely meshed than that Darwin had used. It soon occurred to him that dipping the net would be considerably less laborious than dipping the water, and at last, that towing the net would be even easier. The catch amazed him. Here was a whole world of minute, drifting life whose existence biologists had hardly suspected.

Hundreds of scientists were soon using what came to be called Müller nets (rather unfairly to Darwin, and to other naturalists who had used townets even earlier); the oozy mess they brought up teemed with hundreds of new species of animals and plants. The latter included the diatoms and dinoflagellates, which together make up the bulk of marine plant life.*

The total number of microscopic plants is almost beyond belief. A fourteen-year study near the Isle of Man in Great Britain established that at the most productive season a single cubic foot of water contained twenty thousand or more of them. To be sure, these waters are more productive than the average, though by no means the most fertile in the world. But even if we assume that the average for the ocean's entire lighted zone is only

---

* Some dinoflagellates are not plants but animals that feed on plants—sometimes including their vegetable kin; still others seem to be able to function as either plant or animal depending on the circumstances.

one one-thousandth of this figure, the total number of marine plants alive at any one moment would amount to approximately twenty-five billion *billion*.

In the early 1900's another German biologist, Hans Lohmann, published a study of the *Oikopleura*—a tiny and very simple relative of the vertebrates. This remarkable creature builds itself a gelatinous "house," complete with an emergency exit. Its most important architectural feature, however, is an elaborate feeding apparatus—a tubular structure in which water, flowing through a sort of protective grid, passes through two filtering nets, which collect the organisms that the creature lives on.

Studying this singular residence under the microscope, Lohmann quickly saw that the meshes of the filters were much finer than even the finest silk townet. Yet they obviously must be filtering *something*. To discover what that something was, Lohmann used a centrifuge, the machine with which scientists separate bacteria and other cells from a solution. He discovered that sea water contained even tinier plants than those Müller and his successors had found. Later investigators, using more elaborate methods, have estimated that these minute specks of life are a thousand times as numerous as the teeming diatoms and dinoflagellates that can be caught in a townet. It appears that the oceans as a whole produce about 150 billion tons of plant life a year.

The overwhelming predominance of the very simplest forms among marine plants reflects the relatively uncomplicated conditions of life in the sea. Land plants must have roots to obtain water and minerals from the soil below and branches and leaves to absorb oxygen, carbon dioxide, and energy from the air and sun above. There must be an apparatus to connect the leaves to the roots and transport chemicals and water back and forth between them, some sort of stiffening, so that the aboveground parts of the plant do not simply collapse under the pull of gravity, and finally, a covering to prevent the vital water from evaporating into the air.

Marine plants, on the other hand, are constantly bathed in water that contains the necessities of life, all of which need only be absorbed through the plant's surface into its interior. Therefore, there is a high premium on a large surface area relative to the plant's interior volume. And the simplest way of increasing the surface-to-volume ratio is to be small. If we take a one-inch cube and cut it into eight half-inch cubes, the total volume will remain the same—but the total surface area will jump 50 per cent. And if we repeat this process ten or a dozen times, we will end up with a host of microscopic cubes whose size, and enormous surface-to-volume ratio, approximate those of the larger marine plants.

The smaller the plants, the smaller the plant-eaters. Thus, on land the largest animals are plant-eaters—both now (the elephants) and in the past (the much larger herbivorous dinosaurs)—but at sea the largest are without exception carnivorous, while the herbivores range from small to minute. The only exceptions to this rule are the few species of manatees and the dugong, which live close to shore where they can browse on the ocean's sparse nonmicroscopic plants, the seaweeds and turtle grasses. (The marine turtles, though they evidently thrive on turtle grass, are not exclusively vegetarian; at least two species have been observed munching their way through shoals of jellyfish, seemingly untroubled by those animals' poisonous stingers.)

Thus the great majority of the ocean's strict vegetarians are invisible to the naked eye, like the protozoans, or barely visible, like the copepods or oikopleura. Most of the larger plant-eaters make no distinction between plants and animals, absorbing whatever comes to their collecting apparatus — strainers or threads of mucus — whether it be diatom, protozoan, larval starfish, or nourishing fish egg.

For the same reason, "food chains" — the sequences in which energy is transferred from plants to plant-eaters to eaters of plant-eaters and so on — are likely to be longer in the sea than on land. There are, indeed, plenty of elaborate terrestrial food chains: a butterfly, feeding on plant nectar, may be eaten by a dragonfly, which in turn is gulped by a frog, which is consumed by a snake, which is seized by a hawk. But at least as common is the very simple sequence of plant-herbivore-carnivore, as when leaves or grass is eaten by a deer or an antelope, which is eaten by a wolf or a lion. At sea, however, even the simpler chains are likely to have four or five links — for example, diatom-copepod-herring-tuna-shark.

The transfer of energy up the food chain from eaten to eater involves heavy losses. At a rough estimate, every animal uses up to nine-tenths of the energy it obtains from food simply to stay alive, leaving only a tenth available for incorporation into its own growing tissues—and thus for the nourishment of the next consumer up the line. In the rather oversimplified example above, a thousand-pound shark is produced by consuming some ten thousand pounds of tuna, which represents fifty tons of herring, which represents five hundred tons of copepods, which represents five thousand tons—ten million pounds—of diatoms.

An interesting variation on the marine food chains occurs in regions of strong upwelling, such as that off the Peruvian coast. There, for reasons still not wholly clear, the average size of marine plants is considerably larger than elsewhere; moreover, many microscopic species clump together in colonies large enough to be seen by the naked eye. The result has been the evolution of fish species (the anchoveta is one of them) that feed directly on marine plants. Such simplified food chains greatly increase the already high productivity of commercially useful fish species in the upwelling zones. It has been estimated that these regions, which account for only about a thousandth of the oceans' total area, produce approximately half of the annual world crop of fish.

On the whole, though, of the marine animals large enough to be economically interesting to man, most are fishes three or more links removed from the primary source of energy: even a three-pound mackerel represents about thirty thousand pounds of plant life. It is significant that the largest marine animal, the blue whale, has evolved an apparatus for bypassing several links of the chain. Its whalebone sieve, similar in principle to the far smaller sieve of the oikopleura, enables it to "graze" on small crustaceans that feed directly on plants; it gulps these by the ton as it ploughs through the fertile surface waters off Antarctica. The same is true of the largest fish, the whale shark, which may reach sixty feet or more in length, but—because it is equipped only to feed on small animals strained from the water—is as harmless to man as most of its relatives are inimical.

For marine plants to remain the foundation of the "pyramid of life," they must stay within the ocean's sunlit zone. This is not a great problem for the dinoflagellates; by lashing their flagellums—thin, whiplike appendages—they can, and do, counteract the pull of gravity and propel themselves scores of feet up (or down), presumably drawn by more favorable conditions of light or nutrients. But the diatoms are another story, and a puzzling one.

As mentioned earlier, the smallest marine organisms sink very slowly, but they do sink. That dead diatoms fall to the bottom is proved by the immense stretches of ocean floor covered with "diatomaceous ooze" made of their tiny skeletons. But how do they avoid falling while they are alive?

Many diatoms—as well as many larger organisms—possess elaborate and sometimes grotesque spines or feathery limbs, which undoubtedly retard their rate of fall. Moreover, it has been observed that these appendages, even in members of the same species, will often be longer in summer and in warm water than in winter and in cold water—presumably because warm water, being less dense than cold, requires more of a "parachute" to retard falling. But even the largest parachute cannot indefinitely prevent its wearer from drifting to earth. Evidently, then, the diatoms have some mechanism for making themselves less dense, or at least no denser, than sea water.

Until fairly recently it was assumed that many species of diatoms secreted tiny globules of oil, which, being less dense than water, buoyed them up. Many marine biologists, however, have come to doubt this; the Canadian scientist John D. H. Strickland has gone so far as to call it a myth. A few species are known to produce minute gas bubbles, which apparently serve as floats; others are thought to selectively concentrate certain light substances within their bodies. But the problem as a whole is far from being solved; dissection or refined chemical analysis of a diatom, which may be no more than a few ten-thousandths of an inch long, is a most difficult task.

The problem of defying gravity, of course, confronts all sea organisms except those living on or in the bottom. The solutions they have evolved are diverse; not too surprisingly, most of them are paralleled by the ways devised much later by man to navigate under the sea.

An obvious answer is to carry buoyancy tanks filled with air, as submarines do. Many fish have buoyant air bladders, and a few species of mollusks are kept afloat by air-filled shells. But the air bladder has disadvantages. Its walls (unlike those of the submarine's buoyancy tanks) are not rigid, so if the animal dives, the increasing water pressure will compress the bladder, thereby reducing the animal's buoyancy and perhaps forcing it even deeper, to a level where it will be crushed. If, on the other hand, the fish moves upward, the bladder will expand, quite possibly to the point of fatally damaging the animal's other internal organs.

A fish moves up and down as a submarine does, by "valving" gases dissolved in its bloodstream into or out of its air bladder. But for a fish, unlike a submarine, this is a rather slow process, so air-bladder fish must change depth slowly.

A more elegant solution has been evolved by the cuttlefish, a relative of the squid and the octopus. Its buoyancy tank is a mass of "bone" — actually, as canary fanciers know, a substance much lighter and more porous than ordinary bone. The pores are filled both with gas and with relatively fresh water, which the cuttlefish produces by chemically "pumping" salts out of the pore fluid. Because of a physical phenomenon known as osmotic pressure, the water tends to follow the salts — leaving spaces that fill with gas presumably diffused out of the animal's body fluids. Since the cuttlefish's buoyancy chamber, like that of a submarine, has rigid walls, the animal is not troubled by expansion or contraction of the chamber as it navigates up or down. Moreover, "pumping" salts into or out of the bone pores requires a good deal less metabolic energy than pumping air as fish must do. The only drawback is that the cuttlefish's buoyancy chamber ceases to function at depths below about eight hundred feet.

A few animals resemble the bathyscaph, a deep-sea

exploration vehicle whose buoyancy tank is filled with liquid. Unlike air, all liquids are virtually incompressible. A family of fish called cyclothones, thought by some to be the world's most numerous fish, has bladders filled with oil (the bathyscaph uses gasoline). The squid's tank is water-filled, but the water is permeated with ammonia, which makes it slightly less dense than sea water. The difference in density is so small that the tank must be relatively large—in some species it is twice as big as the rest of the animal. In exchange, however, the device maintains buoyancy at any depth with only a minimal expenditure of energy.

The simplest answer to gravity is simply to keep swimming—which is, in effect, what a moving submarine does when it regulates its depth by the use of its diving planes. This it can do without buoyancy tanks as long as it keeps moving; if its propeller stopped, it would drift to the bottom. In the same way, fish without air bladders, such as the mackerel, its relative, the tuna, and the entire group of sharks, must keep swimming from the moment they hatch until the moment they die. It is noteworthy, however, that most of these perpetual swimmers have a relatively high percentage of oil in their flesh, which lightens their specific gravity.

The same solution of constant activity has been adopted by most of the smaller marine animals that are included among the plankton, or drifters, because of their feeble swimming ability. But feeble though they may be, most of them can move up or down to maintain themselves at a suitable level.

The British marine biologists Sir Alister Hardy and

Richard Bainbridge have actually clocked the swimming speeds of many small planktonic organisms. They employed a device called a plankton wheel, which resembles the revolving "squirrel cages" often used to exercise small pets. The plankton wheel is a transparent plastic, ring-shaped tube, set vertically on an axle and filled with water. As the plankton swim up or down, the wheel is revolved to keep the animals at eye level; a recording device keeps track of the distance covered in a given time.

Many small organisms are remarkably speedy. *Calanus,* one of the most common copepods in temperate waters, can swim upward, against gravity, at the rate of 50 feet an hour. In proportion to its size, which is about that of a grain of rice, this is equivalent to a six-foot man climbing a precipitous hill at a brisk walk. Over short periods *Calanus* can climb at the rate of 200 feet an hour and dive, with the help of gravity, at a speed of up to 331 feet per hour.

Besides the problem of buoyancy, many marine organisms dwelling near the surface must cope with ocean currents. For example, the Gulf Stream, traveling at a rate of three knots or more, can in a matter of weeks transport its freight of marine organisms from the balmy waters off Florida to the chilly Norwegian Sea or to the almost equally cool sea off Portugal and North Africa. In either case the organism at some point will almost certainly face conditions inimical to its survival.

If a fish or a squid finds itself entering waters too cold for comfort, it can turn back and swim toward warmer climes. But the smaller and feebler swimmers, not to mention the completely passive drifters like the diatoms and jellyfish, have no such recourse. The fifty-feet-per-hour speed of *Calanus,* while perfectly adequate for daily migrations of a few hundred feet up and down, is far too slow to buck even a sluggish surface current moving at fifty feet a minute.

The inescapable conclusion seems to be that a great many organisms that are swept into seas too cold or too warm for survival must perish. But what mechanism ensures that there *will* be plenty more where they came from? As the oceanographer R. E. Coker trenchantly puts it, "if everyone in Florida moved to the Carolinas, then on to Virginia, New Jersey, etc., who would start the new migrant populations in Florida?"

One possible explanation is that a "seed stock" of organisms may exist in the eddies or backwaters of the great currents, continually drifting into the mainstream to start a new population. Another theory is that the organisms swept away by the currents either do not die or do not die without issue. G. A. Riley of Yale University has pointed out that in some coastal waters microscopic organisms are known to produce, or to transform themselves into, sporelike forms that can survive very unfavorable conditions and then change back into "normal" organisms when times are better. Moreover, the spores are frequently unrecognizable for what they are; under the microscope there may be no detectable difference between a spore and a speck of dust. Perhaps, then, almost any waters may contain a sort of potential population of far more organisms than are known to inhabit them, with changing conditions determining which creatures will shift from "potential" to "actual."

There is a clear and consistent relationship between mobility and biological complexity. The organisms that are most mobile, such as the fish, the squids, and the crabs and lobsters, have by far the best-developed sense organs and intelligence — or, if intelligence seems too strong a term, at least the capacity for varying responses to varying situations. A codfish or a rock lobster is not very bright in comparison with a man, or even with a mouse — but it is a great deal smarter than an oyster.

The reason, of course, is that the oyster would be unable to make use of the lobster's eyes or the fish's nervous system even if it had them. Immobilized for its entire adult life, its repertory of possible responses to the external world is pretty much limited by its anatomy to opening and closing its shell. And since organisms do not evolve complex organs unless those organs give them a biological advantage, the oyster has remained for millions of years simple and delectable.

Mobile or immobile, simple or complex, all marine organisms must reproduce. And as the example that began this chapter suggests, most of them rely on a prolific production of eggs to ensure the survival and dissemination of their species. Depending on their size, they produce eggs by the thousands or the millions. A possible world's record is held by the ling, a relative of the cod, a 54-pound representative of which was found to contain more than twenty-eight million eggs.* All marine animals begin as free-floating larvae,

---

* Recent studies of the sea hare, a kind of mollusk, suggest that it may outdo even the ling in fecundity; on occasion it may lay more than 80 million eggs at a clip.

even those that spend most of their lives fixed in one place, like the oysters and the barnacles, or those that roam only a limited area of the bottom, like the starfish and the worms. Indeed, these larvae make up a very respectable proportion of the plankton, and as such, share the fate of the other planktonic organisms, which is almost invariably to be eaten. For this reason, marine animals must maintain a prodigious rate of reproduction if even a few larvae are to survive into adulthood.

Some animals living in more or less protected habitats have forgone prodigality by developing ways of protecting their young until they are big enough to have a somewhat better chance of survival. The ocean pout, for example, lays her eggs—only a few hundred at a time—in cans, old boots, and similar junk that has made its way to the ocean bottom. Even more remarkable is the sea horse, a common inhabitant of the eelgrass beds. The female deposits her eggs in a pouch under the tail of her spouse, where they are nourished by the father's bloodstream. They emerge as almost fully formed sea horses rather than as eggs or larvae, by a process that bears a rather startling superficial resemblance to human labor.

Jack Rudloe, a zoologist who collects marine animals for aquariums and research scientists, has described one such male sea horse whose labor pangs he witnessed in one of his collecting tanks:

He was swimming erratically around the tank, bumping into the glass, then dashing back to the other end. . . . Suddenly the big sea horse spun around and raced back to his favorite perch, wrapped his tail around it, contracted hard and convulsively and sprayed out what seemed a multitude of tiny black sea horses. The infants, no more than a quarter of an inch, floated up to the top of the water. . . . Three times he went through the birth agonies, presenting me with two hundred and eleven babies.

But the sea horse, with its relatively modest output of a couple of hundred young, is one of the exceptions. As a rule, sea creatures seek biological safety in numbers. Because their reproduction requires a prodigious expenditure of energy, marine animals (like nearly all land animals, with the outstanding exception of man) reproduce only at certain times — once a year in the coldest waters, several times a year in warmer regions. For a wide variety of species, including many fish, this necessitates some sort of arrangement for bringing males and females together at the appropriate times. And for all species, both sexes must be fertile when they meet.

The salmon hatches in fresh water and swims downstream to the ocean, where it remains until it is sexually mature. Then both sexes return to their native stream to spawn and die. Their return is triggered by the readiness of their reproductive organs for spawning.

How they find their way back to the river mouth, and even the tributary brook, where they were born has been the subject of much research, largely due to the economic importance of the Pacific Coast salmon fisheries. At one time the salmon was credited with almost supernatural navigating ability, but more recently some scientists have come to doubt this. A researcher at the University of Rhode Island has suggested that salmon may swim rather randomly in the ocean — but are so numerous that enough of them to maintain the population of a given river system will make their way from the open sea to coastal waters at the mouth of the parent river. Computer calculations indicate that the numbers of salmon returning by sheer chance should square pretty well with the actual numbers that do return.

Once in the river, however, the salmon almost certainly picks out his native stream from twenty other tributaries not by chance but by means of a chemical sense akin to a combination of smell and taste. In one study salmon were netted in their native stream and then returned to the lower river with their nostrils plugged; none of them could find their way home again, though salmon with unplugged nostrils had no difficulty. Evidently there are minute differences in the chemistry of streams even a mile or two apart — far too subtle for human detection, but clear enough to a salmon.

Many wholly oceanic fish return as regularly to their own spawning grounds as the salmon does to its native stream — quite probably guided by similar chemical attractions. There are, for example, three known eel-spawning grounds in addition to the Sargasso Sea: one in the North Pacific, which supplies eels to Japan and China, another in the South Pacific, used by the eel population of Australia and New Zealand, and the last in the Indian Ocean. Cod spawn by preference in waters around thirty fathoms deep; the submarine shoals, or "banks," where this occurs, off Norway, Iceland, Greenland, and Newfoundland, have, not surprisingly, become major cod fisheries.

In the depths of the ocean, where light does not penetrate and living creatures are thinly scattered, the problem of ensuring that boy meets girl is rather more difficult. We know very little about the reproductive activities of marine life in these regions, although it seems likely that the patterned luminous organs possessed by many deepwater species serve in part to bring the sexes together.

A more radical solution has been adopted by the deepwater anglerfish. The males attach themselves at an early age to the body of the females, living out their lives as parasites on their mates. Their only function is to fertilize the eggs when they appear—a rather discouraging commentary on the biological importance of the male animal.

There seems little doubt that some deepwater fish find their mates through sound communication—as some shallow-water fish certainly do. Captain Jacques Cousteau, the inventor of scuba diving, entitled his first book on submarine exploration *The Silent World*, but the ocean has turned out to be, in fact, a rather noisy place. World War II submariners discovered this when their hydrophones picked up mysterious beeps, groans, crackles, and whistles—sounds that could not originate from any ship since no ships were around. Postwar investigations established that the sounds were produced by marine animals—notably fish, porpoises, whales, and shrimp. The sea robin, a particularly noisy species, vibrates the walls of its air bladder, producing a chorus of grunts and cackles. The triggerfish has modified part of its air bladder into a drumlike membrane on which it beats with its pectoral fins; the squirrelfish "grinds together toothed areas in the back of its mouth . . . and thus generates a sound which is amplified by the adjacent air bladder and becomes a rasping grunt." And shoals of shrimp, ejecting jets of water for defense or offense, produce a crackling like "fat frying in a pan." Porpoises, as we shall see later, use their repertory of sounds not only to communicate but also as sonar signals to range in on an approaching meal.

A marine biologist with the appropriate name of Marie Poland Fish has described the "boat whistle" mating call of the toadfish, a coastal species that is quite as ugly as its name suggests: "The calls of individual fish were repeated at regular intervals of about 30 seconds, and sometimes we heard what seemed to be an answer from some fish at a distance. When the females had deposited their eggs in the shallow inlets, these clear loud calls ended, and afterward the only sound that could be evoked was the low, coarse growl of the pugnacious males guarding the nests."

But the bringing together of the sexes, through whatever means, will not by itself ensure that the males will be ready to fertilize the eggs when the females are ready to lay them. In a number of species the matter of synchronizing reproductive readiness seems to be controlled by what are called biological clocks.

Scientists have long known that a great many animals — and plants — show regular changes in their physiology and activities that are synchronized with the cycles of day and night; thus, many plants flower at specific times during each twenty-four hours, presumably to ensure the attentions of the equally cyclical insects that pollinate the blooms. The naturalist Linnaeus once planted a bed of flowers chosen so that the opening of a particular bloom would mark each hour of the day and night; he called this array "Flora's clocks." In temperate regions plants flower and bear fruit, and animals and birds migrate, in a seasonal rhythm geared to the changing length of daylight during spring, summer, and autumn.

In addition to daily and seasonal cycles, many sea organisms change according to the phases of the moon, which in turn control the tides. The grunion, a small fish inhabiting California coastal waters, apparently has just such a lunar cycle of reproductive ripeness. A night or two after a full or a new moon, which mark the highest tides of the month, both sexes swim to the beaches; the females lay their eggs in the sand near the high-water mark, and the males fertilize them. This biweekly cycle also ensures that the eggs will be able to hatch and develop without interference for nearly two weeks, when the next peak in the tides will roll up the beach and sweep the young out to sea.

A marine worm native to West Indian waters is similarly influenced by the lunar cycle. Although normally living in the bottom, it swims to the surface to breed about the time the moon is in its third quarter. The females emit their eggs along with a stream of brightly luminous fluid; this attracts the males, who eject their sperm over the eggs, producing shorter flashes of light.

There is a historical footnote attached to this particular worm. On the night of October 11, 1492, the half-mutinous crew of Columbus's *Santa Maria* were startled when the lookout spied a light. Historians have taken this to be an Indian campfire, but there is some doubt as to whether the ship was then close enough to land for a campfire to be visible. Naturalists have pointed out, however, that the ship was certainly in waters where this worm is known to conduct its luminous love-making. And October 11 was only one night before the moon's third quarter. . . .

Some lunar sexual cycles seem to be governed directly by the monthly variations in the moon's light, from dark of the moon to full; others seem to result from interaction between a daily and a tidal cycle. The "tidal day" is about fifty minutes longer than the solar day, which means that high (or low) tides gradually shift their timing with relation to the sun. Thus, for example, the mating activity of a particular species might be set off only when high tide happens to fall at sunset; this would occur twice a month at most, and in some areas (where there is only one high tide a day) just once a month. There is evidence that some marine animals are also subject to an annual sexual cycle (as many land animals are) and that the combination of the three cycles ensures that nearly all the individuals in a given locality will spawn during only a few hours out of the entire year.

Such an animal is the palolo, or mbalolo, worm of the western Pacific. Its various internal "clocks" are so arranged that spawning occurs only in October or

November and only when high tide occurs at dawn. The results of this mass mating are spectacular. The worm, an inhabitant of the coral reefs, reproduces by detaching the entire back half of its body, which rises to the surface and emits the eggs or sperms. We owe the classic description of this phenomenon to one of the many British officials who, to relieve the tedium of their posts in distant colonies, wrote down their observations of the curious creatures, human and otherwise, which they found around them.

"When the first light of dawn appears," the official wrote, "great funnels of worms burst to the surface and spread out until the whole area is a wriggling mass of them, brown and green in color. When the tropical sun rises perpendicularly from the sea, the catch is in full swing, and hundreds of boats, canoes and punts are filling up kerosene tins and jars." The palolo, he added, "is rightly prized as very good eating."

That "rightly"—evidently based on a personal experiment that few of us would have dared to undertake—points up an important aspect of life in the sea: the participation of our own species in its incessant cycle of eat-and-be-eaten. The palolo and hundreds of other marine species contribute to man's diet around the world, which will be discussed in the next chapter. And—as we shall see in Chapter V—man on occasion finds himself in the opposite situation, and is eaten in turn.

CHAPTER IV

LOTS OF GOOD FISH

When it comes to food, Homo sapiens is about as unchoosy a species as you are likely to find. At various times and places his menu has included caterpillars, termites, snakes, lizards, the "100-year-old-eggs" and birds'-nest soup of Chinese gourmets, and the many types of fermented, moldy, or rancid milk products that tempt the European or American cheese fancier.

Though there is no master list of marine organisms that have made their way onto our dinner tables, the sea's contribution is hardly less diversified than the land's. The Japanese gather nearly a dozen species of edible seaweeds, and some American "Down Easters" relish a species called dulse; still another seaweed, Irish moss, is used as a thickening agent for puddings. Sea urchins are esteemed a delicacy by many Frenchmen, as are octopus and squid by people all around the Mediterranean. In the South Seas gutted sea cucumbers are dried and sold as *bêche-de-mer*. The Eskimo enjoys his seal and walrus blubber while sea turtles make their way into the cooking pots of Caribbean Indians and the ceremonial tureens of London aldermen.

The great bulk of the world's seafood, however, is

drawn from the three groups of animals that account for practically all the marine organisms of any size, *i.e.*, that are worth the trouble of gathering. These are the crustaceans (shrimp, lobsters, crabs), the mollusks (shellfish of all sorts, in addition to squid and octopus), and above all, the fish.

We do not know precisely when man—who is, after all, a land animal—began to forage in salt water. It is not unlikely that at least a few bands of apelike subhumans lived along the sea coast several hundred thousand years ago, or even earlier. But the evidence of whether they fed on marine animals has been obliterated by subsequent changes in sea level produced by the periodic locking up of millions of cubic miles of water in the great continental glaciers.

Deposits in a Middle Eastern cave reveal that some forty thousand years ago its Neanderthal tenants ate land snails, and we can thereby infer that others were very likely eating sea snails and other shellfish. Certainly there is little doubt that mollusks were the first marine addition to man's diet, for gathering them requires neither great dexterity nor specialized equipment. Nearly all shellfish are either immobile (oysters and mussels) or move at a snail's (or a clam's) pace. Near my summer home on Cape Cod there is a sandbank where, when the tide is low enough, I can gather enough clams for a mammoth chowder in five minutes, using only my feet (to feel out their location) and my hands (to dig).

A bit of reindeer horn some fifteen thousand years old, dug up in southwest France, is engraved with a vigorous representation of leaping salmon, and again we can infer that if some men of those days caught salmon in rivers, others may well have caught different kinds of fish in estuaries and bays.

The first concrete evidence of man as a fisherman, however, dates from eight to ten thousand years ago. By that time, people in Europe and elsewhere were using harpoons, fishhooks, and nets to harvest a variety of marine animals. In many places along the coast of Europe and North America the sites of villages can be located by enormous heaps of shells and fish bones, which archaeologists delicately describe as "kitchen middens," but which were in fact garbage dumps — and fragrant places they must have been, too.

Since then, fish and fisheries have played a not inconsiderable role in human culture and at times have even influenced the course of history. The Cretans of thirty-five hundred years ago decorated their pots and vases with lively depictions of fish, squid, and octopus, no doubt reflecting a culinary interest still shared by inhabitants of the Mediterranean lands. In the year A.D. 647 Felix, bishop of the East Angles, considered the North Sea herring fisheries important enough to establish a church with "a godly man placed in it to pray for the health and success of the fishermen that came to fish at Yarmouth in the herring season." (Yarmouth is still an important North Sea fishing port.)

During the Middle Ages the great herring fisheries of the Baltic helped launch the trading empire of the Hanseatic League. And the decline of those fisheries during the fifteenth century crippled the economy of

North Germany and, so it has been argued, helped launch Germany on the troubled and troublesome career that culminated a generation ago.

Certainly fisheries helped open North America to settlement. Only a few years after Columbus's historic voyage John Cabot explored the region of the Grand Banks and found the waters "swarming with fish which can be taken not only with the net but in baskets let down with a stone, so that it sinks in the water . . ." For a century before the Pilgrim Fathers landed, fishermen from half a dozen nations had been netting or hooking cod on the Banks and other coastal waters from Massachusetts north and carrying them, dried or salted, to hungry Europe. The Pilgrims' first landing, indeed, was made on the great peninsula that was already beginning to be known as Cape Cod. The mercantile wealth of New England was partly based on fishing (and whaling), as witness the carved wooden cod that occupies a place of honor in the Massachusetts State House, and New England clipper crews hoisted sail to the tune of:

Cape Cod girls don't have any combs,
Heave away, heave away!
They comb their hair with the codfish bones
And we're bound for Australia.

For a nation partly founded upon fish bones, the United States today consumes very little seafood. The average American annually eats some 170 pounds of meat and 32 pounds of poultry—but little more than 10 pounds of fish. Yet as a source of protein, the substance that our bodies require for growth and repair, fish is fully equal to flesh or fowl. Nations such as Japan, in fact, obtain up to 80 per cent of their animal protein from the sea. John D. Isaacs of the Scripps Institute of Oceanography estimates that the oceans could supply the entire animal protein needs of several times the world's present population, and while his estimate is almost certainly overoptimistic, it is a fact that fish is one of the very few basic foodstuffs whose production has been increasing *faster* than the world's population.

The nutritional importance of fish stems from the fact that a large proportion of the people in the world — some say as many as half — lack sufficient protein; at least 100 million suffer the acute protein-deficiency disease known as kwashiorkor, which can produce permanent physical and mental damage, and in severe cases, death. To meet this need for protein through stock raising would, in most ill-nourished countries, simply displace other badly needed food crops or lead to overgrazing of the land. Expanded fisheries are an obvious alternative.

Better technology is a typical American response to the problem of increasing the harvest from the sea. Its advocates point out that the basic tools of fishing — nets and hooks — have not changed in any fundamental way for thousands of years. Why not, they ask, put modern engineering research to work and devise ways of fishing that are as up to date as those now used in American agriculture, which is unquestionably the most mechanized and efficient in the world? Underwater "curtains" of dyes or air bubbles might be used to round up schools of fish; underwater television could guide trawl nets to the densest shoals, or perhaps the trawl itself could be attached to a manned underwater vehicle that would pursue the fish in their own element. Another method of collecting fish would employ electric currents — which already are being used by some fishermen to concentrate netted fish for more effective transfer to their ships.

A number of scientists, noting that salmon — and very likely, other fish — find their way to their spawning grounds by detecting chemical "odors" in the water, have proposed that by identifying these substances and scattering them in parts of the ocean we could lead the fish to the net. Perhaps the farthest out proposal has been made by Marie Poland Fish, the expert on marine sounds introduced in the preceding chapter; she suggests that fish can be "herded" by underwater speakers that emit alarming, or attracting, fish noises.

Other experts, however, are by no means sure that gadgetry of this sort is the whole answer, or even an important part of it. They point out that today's relatively primitive methods are quite efficient enough to overexploit some existing fisheries. The California sardine industry, for example, was all but wiped out in the late 1940's and early 1950's. Overfishing, and a drop in sea temperatures that cut the fishes' spawning rate, so diminished the sardine population that the fisheries have never recovered. The sardines' place has been taken by anchovies, which have little commercial value at present. As the best fishing grounds are fished more intensely, it is becoming clear that international regulation of the catch is the only way to ensure that other valuable fish populations do not go the way of the California sardine.

A crude sort of regulation has been going on for generations. Decreases in the abundance of certain fish—which now seem fairly certain to have been caused by temporary natural factors—long ago led many of our state governments to restrict the size of boats, the type of gear, or the length of nets that might be used. But such measures, lacking any scientific basis, served chiefly to perpetuate inefficient techniques (and sometimes to attract more fishermen into an already overcrowded field) rather than to conserve the fish supply. Moreover, local regulations cannot control fishermen of other nations.

A more productive approach was to regulate—often by international convention—the mesh size of nets allowed in a given fishery, as has occurred in several important North Sea fishing grounds. The reasoning was that larger meshes would permit immature fish to escape, to be caught another year when they were bigger. At first some fishermen (and scientists) argued that regulating the mesh size would often be pointless, since in many types of fishing gear—notably the drag or trawl net—the fish were so crowded together that even small fish could not slip through the meshes. But underwater photographs taken by frogmen proved that at least some small fish could and did swim out of the trawl. More recently, however, fisheries' scientists have established that net size alone is not the answer. No matter how faithfully the regulations are observed, a sizable number of immature fish will still be caught—

enough, under conditions of intensive fishing, to seriously deplete future stocks.

The imposition of a closed season on certain fish has also proved temporarily effective. An agreement of this sort between the United States and Canada, put into effect in 1932, raised the annual catch of halibut in the northeast Pacific from forty-two million pounds to its present seventy-five million pounds. However, limiting the fishing season takes into account neither the number of boats fishing nor year-to-year variations in the fish "crop" — that is, the number reaching commercial size each year.

The only rational basis for conserving fish — or any other self-renewing natural resource — is to determine the "maximum sustainable yield" of a given fishing ground and divide this among the various nations using the ground according to some quota system. A complicating factor, however, is that, for reasons we know relatively little about, fish populations vary markedly from year to year in a given region.

Oddly enough, the size of each year's fish crop seems to have relatively little to do with the number of eggs or of spawning adults. Apparently, an especially large crop of eggs usually means, other things being equal, merely that a larger proportion of them will be eaten. With a small crop, on the other hand, the eggs are so scattered through the water that the rate of attrition will drop from, say, 99.99 per cent to 99.98 per cent — meaning that twice as many eggs per million will survive to adulthood.

There is a good deal of evidence to back up this

theory. Some of the earliest findings come from the work of the great Norwegian fisheries expert Einer Lea, whose monumental study of "age composition" in the Norwegian herring fishery is typical of a great deal of fisheries research — and, for that matter, a great deal of oceanographic research generally. It involved none of the sensational discoveries that occur in the history of physics and chemistry. Rather, it consisted of innumerable bits of evidence, their careful examination and classification, and ultimately, their interpretation in graphs showing the rise and fall of the varying age groups in the fish population.

Lea's evidence was herring scales. Under a magnifying glass they show a pattern of rings that Lea believed, and ultimately proved, were like those of a tree — annual growth rings from which the age of the fish could be deduced.

For seventeen years, from 1907 to 1923, Lea, with the help of fishing-boat captains, collected tens of thousands of scales from a sampling of the herring landed in Norwegian ports, classified them by age, and entered the results in lengthy tables. Soon he realized that something very extraordinary had happened to the herring spawned in the year 1904. As early as 1908 these four-year-old fish accounted for a greater percentage of the catch than any other age group; by 1910 they made up nearly 80 per cent of the herring landed, and even thirteen years later, at the very advanced age of eighteen years, the 1904 herring still accounted for more of the catch than many younger groups.

One might think that this bumper crop of 1904

herring must have resulted from a correspondingly large number of adult herring spawning in that year. Since Lea had not been collecting scales in 1904, he could not answer the question directly, but a close examination of his tables shows that the herring five to ten years older than the 1904 group—those old enough to account for the bulk of the eggs laid in 1904—were, as a group, not especially numerous. The only explanation seems to be that far more of the 1904 eggs survived to become adult herring.

Some clues to the possible reasons have come from the work of Sir Alister Hardy, whom we have already met as a clocker of swimming plankton. His first research project as a young marine biologist was to piece together a picture of the herring's diet at the various stages of its life, from a hatchling less than half an inch long to an adult of a foot or more. His technique was hardly less laborious than Lea's. He selected herrings caught by fishermen and baby herrings of various sizes hauled up in plankton nets, dissected them, and examined the contents of their guts, under the microscope when necessary, to determine what they had been feeding on.

The complex conclusion he came to is worth reviewing briefly to correct the deliberately oversimplified picture of marine food chains given earlier. A marine organism does not consistently eat the same thing all its life (as some land animals do). Not infrequently, a species that early in life is preyed on by some other animal will in its adult years prey on that same predator; an adult herring, for example, may well consume an arrowworm that fattened on its own herring hatch-lings.

The newly hatched herring is a vegetarian, feeding on diatoms and flagellates, but it quickly graduates to a "meat" diet of very small crustaceans and the larvae of larger crustaceans, mollusks, and other creatures. As an adult the herring feeds by preference on the copepod *Calanus*, but lacking that, it will snap at many other small organisms, including the sand eel (not an eel but a slender fish). Hardy has described how the stomach of a herring will often be found crammed full of sand eels, "lying neatly side by side like sardines in a tin."

An important finding, Hardy believes, is that for about a month of its early life the herring—at least in the North Sea—appears to feed almost entirely on a single species of crustacean, the tiny copepod *Pseudocalanus elongatus*. Obviously if there happens to be a shortage of this particular creature during the crucial month, "there may be a heavy mortality among the baby herring and this may deplete the stocks of adult fish in later years." Conversely, of course, an especially abundant crop of *Pseudocalanus* one year may mean a bumper crop of adult herring five years later.

Alternatively, a bumper crop of predators may drastically reduce the number of young herring. The sea gooseberry, which looks rather like a small jellyfish but isn't one at all, can eat as many as five baby herring at once, and an explosive multiplication of these animals could wreak havoc on the fisheries later on.

It would seem that the number of adult fish at any given time depends less on the number of eggs laid than on the presence or absence of food and enemies at various times during their growing up. Lately, however, fisheries' biologists have begun to wonder whether the sort of intensive fishing now going on might not reduce the breeding stock to levels that could cripple the fish harvest for years or even decades—the more so in that some other factor may intensify the effects of overfishing, as was the case with the anchovies that displaced the dwindling shoals of California sardines. We shall have more to say later about the ways in which the commercial-fisheries people and other researchers are trying to put the harvesting of fish onto a firmer scientific footing.

One method of increasing the world's fish output that requires relatively little research and involves few risks is to harvest more species. Senator Claiborne Pell, like most Rhode Islanders, is intensely aware of the sea and of the problems of fisheries and fishermen. He points out, for example, that consumer demand for a particular species of fish is often just a matter of marketing. Thus the redfish, a common species along our North Atlantic coast, was little used until some shrewd individual thought to rename it "ocean perch"; today, tons of frozen "ocean perch" fillets are sold each year. Again, the deepwater red crab is so plentiful as to be a positive nuisance to lobster trawlers, at times forcing them to haul nets and fish elsewhere. Yet the crab, now commercially worthless, contains more meat per pound than the blue crab, which furnishes us with such gourmet dishes as crab cakes Maryland and fried soft-shelled crab; some prefer the red crab's meat to lobster.

Fishermen off Point Judith, Rhode Island, says the senator, have taken one and a half to two *tons* of red crab in an hour's trawling; to gather an equal quantity of lobster requires two days and nights at sea. Development of a machine for picking out the meat of the red crab (such as the one that already exists for extracting the meat of blue crabs) could convert a nuisance into a valuable resource—and incidentally might take some of the pressure off the lobster, which, thanks to the unprecedented demand generated by American affluence, becomes rarer and more expensive each year.

Nearly half the fish netted by American boats are immediately thrown back into the sea as "trash." But from a nutritional standpoint all fish contain equivalent protein; the only problem is how to get that protein into the stomachs that need it.

One promising approach is to convert fish of any species into what is called fish protein concentrate (FPC). This substance is produced by grinding up whole fish, dehydrating the pulp, and then chemically removing oils and odorous substances. The resulting bland powder, with a taste resembling malted milk, can be mixed with flour or used to fortify soups, stews, and so on. It costs perhaps thirty cents a pound, and since it contains four times as much protein as fresh meat, it is nutritionally equivalent to meat at about seven and a half cents a pound!

Whether FPC will be accepted among people now suffering from protein deficiency is still to be determined. Unfortunately, its development was held up for some years by one of those bureaucratic hassles that seem to be endemic in large organizations, governmental or private.

The Food and Drug Administration had to approve FPC before it could be marketed for human consumption in the United States. But the FDA declared that the concentrate was "aesthetically" undesirable because it included bones, scales, and guts, and moreover, it violated FDA regulations against food containing a "filthy, decomposed or putrid substance."

The makers of FPC were not unduly discouraged by this decision because they had not anticipated any large market for their product in meat-eating, protein-rich America. But another bureau, concerned with America's international image, then forbade the export of FPC, refusing to permit the sale to foreigners of a product declared unfit for American consumption.

Commentators pointed out that in making "aesthetic" judgments the FDA had exceeded its authority; moreover, sardines — which neither the FDA nor consumers had ever found either unaesthetic or "filthy" — were just as much whole fish as FPC. Subsequently, following a change of administrators at FDA, the concentrate was finally approved. Now the question is whether it will find the market it deserves among protein-deficient people.

The problem of increasing the output of marine protein has caused some to suggest harvesting not fish but the plankton they feed on. Superficially, this seems reasonable. Plankton, after all, are at or near the base of the food chain; most fish are close to the top. And if, as we have seen, something like a thousand pounds of plankton is needed to produce one pound of fish, why not harvest the thousand pounds rather than the one?

The flaw in this proposal is that although there is plenty of plankton, it is so thinly scattered that to collect a ton of it would require filtering up to a million tons of sea water, at a prohibitive cost in equipment, fuel, and human energy. There is gold in sea water too, by the millions of tons, but nobody has yet figured out a way of concentrating it at a profit. Far better, say the experts, to leave the harvesting of plankton to fish and other marine organisms naturally equipped for the job. These organisms may be inefficient by human standards—but they work cheap.

The simplest way of increasing the marine harvest is to expand the harvested area, for it is becoming clear that some of the most productive waters are hardly being exploited at all. During the past decade or so Peru has begun for the first time to fish the cold, fertile waters off its coast with efficient, modern equipment. As a result it has become the leader in world fish production, accounting for approximately one-sixth of the total tonnage. (Nearly all the Peruvian harvest is made up of anchovies or anchovetas, which are processed into fish meal to feed livestock.)

As pointed out earlier, the chief physical factor in determining marine productivity is the supply of nutrients, which in turn depends upon either the runoff from the land or the upwelling of deep water. In recent years Ghana and other West African nations have begun to fish the Gulf of Guinea, which is fertilized by the powerful flow of the Niger and the Congo, among other rivers. Still to be intensively exploited is a fishery of possibly equal potential just across the Atlantic, off the mouths of the Amazon.*

The physical oceanographers are learning that upwelling occurs much more commonly than had once been thought (though seldom on the magnificent scale we find off the Peruvian coast). Diverging surface currents, like those found along much of the equator, are accompanied by upwelling, since the "outward bound" surface water must be replaced by water from below. Taking advantage of this discovery, the Japanese have already developed a valuable tuna fishery in the Equatorial Pacific.

Upwelling of a sort has also been found to occur at the boundaries between currents moving in opposite directions. The opposing motions produce turbulent horizontal eddies and also vertical ones. Eddies of this sort between the southwest-flowing Labrador Current and the northeast-flowing Gulf Stream help create the enormous productivity of the Grand Banks. Similar but far less exploited turbulence belts exist, among other places, about 10 degrees north of the equator, where the North Equatorial Current and the Equatorial Countercurrent glide past one another.

Vertical turbulence, and an enhanced nutrient supply, can also be produced by abrupt protrusions from the sea bed—oceanic islands and the submarine mountains known as seamounts and guyots. Some of these, such as Cobb Sea Mount, which was discovered in the

---

* Runoff works both ways, of course. Completion of the Aswan Dam in Egypt has almost eliminated the flow of Nile water into the eastern Mediterranean during some periods of the year. The reduction of incoming nutrients has wreaked severe damage on commercial fisheries in the area.

North Pacific by the U.S. research vessel *John N. Cobb*, have already become popular trawling grounds.

Full exploitation of all the potential fishing grounds will necessitate a good deal more research to accurately map their location, extent, and productivity—all of which can be expected to change somewhat from season to season and often from year to year. When sending a fishing fleet to grounds some thousands of miles away (as now occurs on proven grounds in the North Atlantic and Pacific), nobody can afford to do so on the off-chance that fish *might* be there; they must be reasonably certain that fish *will* be there.

It has been suggested that in certain areas the nutrient-rich deep waters might be lifted artificially to the surface to simulate upwelling. Pumping or similar methods are out of the question because of the energy involved, and a proposal to warm the bottom water with nuclear reactors seems hardly less farfetched. More plausible, perhaps, is the proposal to heat the water with nuclear wastes. Certainly as nuclear power becomes more widespread, disposing of the "hot" wastes produced by reactors will pose increasing problems. By dumping the wastes, sealed in long-lived watertight (but not heat-tight) containers, in likely areas for raising nutrients, it is just possible that we might reap a bonus in the form of increased fish production near the surface.

Can man successfully "farm" the sea? In one sense, the answer is yes. Mussels, clams, and oysters have been farmed for centuries.

During the spawning season at many oyster fisheries,

the hatching oysters, or spat, are caught on trays containing tiles, to which they attach themselves. At a certain age the tiles are transferred to oyster beds offshore, and sometimes the adult oysters are moved again, into protected waters for fattening just before being marketed. The Japanese have gone a step farther: they arrange for the spat to attach themselves to long ropes hung from rafts anchored in protected waters, thereby forming a multilayered oyster bed that can extract food from the whole volume of water between surface and bottom instead of from the bottom water alone. Italians do much the same thing with mussels by encouraging them to attach themselves to sticks thrust into the bottom. The resulting three-dimensional beds can produce more than 100,000 pounds of mussels per acre per year—approximately one hundred times the yield under natural conditions.

Shellfish are very simple to farm because once they are planted they stay put, like crops on land. Farming moving fish in the open sea is a problem that at present seems insoluble. (Fish farming in fresh-water ponds has long been practiced in southeast Asia, and the Japanese are now experimenting with similar farms in shallow bays and estuaries that can be fenced off from the sea.) Efficient farming involves not merely raising food organisms and harvesting them but weeding out competing organisms and predators. The sea is still far too big for such activities. A much more promising approach is to "ranch" rather than farm the sea. One of the first research projects along those lines is now going on at the University of Washington under the leadership of Professor Lauren Donaldson.

The Washington group is engaged in improving the Chinook salmon through selective breeding—a technique that has been practiced, albeit less systematically, since man first began to domesticate plants and animals.*

The habitat of the Chinook salmon is the once wild rivers of the northwestern United States. Because of dams built to generate power and control floods, canals and locks built for navigation, and—not least—pollution caused by expanding populations and new industries, the salmon catch in the Northwest has dropped catastrophically. In 1883 the Columbia River alone yielded some forty million pounds of salmon; today the catch is about five million pounds.

On the assumption that man was not likely to change the rivers, the Donaldson group set about changing the salmon. They transplanted Chinook fingerlings into an experimental pool near Seattle, which is fed by polluted Lake Washington and communicates with the salt water of Puget Sound through the Lake Washington Ship Canal. Many died, of course, and many more failed to return as adults from their journey out into the Pacific. But by selecting the largest and most vigorous of the survivors, and by intensive feeding of their progeny, the researchers have produced fish that not only can survive in the new environment but are far more productive than "wild" salmon.

This has come about in two ways. First, the breeding

---

* For example, the wild corn that grew in Mexico some seven thousand years ago produced cobs an inch or so long, bearing some fifty tiny kernels. After only a few centuries of cultivation it began to show similarities to the far more productive corn of today.

stock has been continuously selected and crossbred for rapid growth, vigor, resistance to disease and parasites, and so on. Second, the hatchlings have been given a specially nutritious diet, chiefly composed of ground-up hake (an ocean fish of relatively small commercial value). The researchers have sensibly assumed that the best balance of nutrients for building up fish tissues should be found in fish tissues. As a result, their hatchlings grow far more rapidly than wild salmon, and they are ready to migrate to sea in five months, rather than the normal twelve. Since the annual "run" of spawning salmon occurs in autumn, the specially fed hatchlings will reach the open sea, not in the following autumn, but in the following spring—which is to say, close to the peak of the annual "bloom" of their plankton food. Moreover, they reach the ocean weighing two to three times as much as normal young salmon, with a correspondingly greater capacity to feed and to evade predators. Finally, the improved salmon mature much faster; normally the Chinook spends four years in the ocean, but some of the new breed return, full-grown and sexually mature, in as little as one year. And the less time they spend in the ocean, the better their chance of returning from its dangers.

The effect of all these factors shows up most dramatically in the rate of return of the adult salmon. Before being allowed to swim out to sea, the hatchlings are permanently marked—like cattle—with a painless, miniature branding iron. When the project started some twenty years ago, only one out of every thousand branded fish made its way back to the home pool, a rate of return comparable to that measured under natural conditions. Today the run of returning adult salmon ranges from ten to thirty-two times that figure.

The Washington group is now beginning to "plant" other tamed rivers in the Northwest with their specially fed, specially bred hatchlings, in the expectation that the run can be similarly stepped up. At present Donaldson does not expect that his "super salmon" will establish themselves as a natural population; each year the stock will have to be replenished from his breeding pool. But the enormous jump in harvestable adult salmon, he believes, will compensate for the expense involved. In principle, there seems to be no reason why selective breeding could not be applied to fish such as the shad and striped bass, which also hatch in rivers, and after growing to adulthood in the ocean, return to their spawning grounds — and the fish market.

Selective breeding of wholly oceanic fish is much farther off, for there is still too much that we don't know about why they do or do not survive. And it is worth noting that none of the highly productive varieties of land plants and animals that man has developed could flourish, or in most cases even survive, without the careful tending they receive from the farmer or stockman. The day we can convert the limitless ocean into a garden has yet to come.

Moreover, the dubious outlook for international cooperation darkens even the present prospects for increasing fish production. The best scientific information is worthless unless men are willing to act on it; the subtlest calculations of the "maximum sustainable yield" of fish stocks are wasted computer time unless all interested nations agree to limit their catches accordingly. Agreement must be reached on national quotas for particular grounds, and an effective international apparatus must be established to make sure that the quotas are observed.

Unfortunately, the past history of fishing regulations on the high seas does not augur well for their future success. In the case of the Antarctic whale fisheries, for example (an appalling story that will be told in Chapter VI), the greed of the whaling nations in setting quotas, plus feeble and at times corrupt enforcement, has led to the virtual destruction of the main commercial whale species. Recovery of whale stocks to normal levels will take at least a generation, and may never occur at all.

Today all governments give lip service to the cause of international co-operation in regulation of the fisheries. But when it comes to appropriating money for the woefully underfinanced international fisheries bodies, they suddenly discover pressing previous commitments. As in so many other areas of conservation, agreement that something needs to be done in no sense guarantees that it will be done.

CHAPTER V
THE BAD ONES

The first man who put to sea beyond sight of land and returned safely undoubtedly told some tremendous tales about what he had seen. Through the ages, descriptions of tempests, whirlpools and waterspouts, and monsters, all suitably inflated, have enlivened the conversation of the sailor home from the sea.

In a cave at the water's edge Ulysses encountered Scylla, who snatched his shipmates away with her fearsome arms. Sinbad told of a fish so big that he mistook it for an island and landed on it. Seafarers of the Middle Ages chilled landlubbers' blood with tales of the kraken, which could pull the stoutest ship down to a watery grave. And sea serpents have undulated through the waves of half a dozen oceans.

During the past century or two scientists have managed to pin down the considerably tamer realities that underlay the hair-raising yarns of the ancient mariners. Sinbad's enormous fish was doubtless a whale, the largest animal on earth—but hardly large enough to be mistaken for even an islet. The kraken and Scylla are thought to have been giant squid. But these fearsome-looking tentacled creatures, though they can reach fifty feet in length, have never been proved to have sunk even a rowboat. As for the sea serpent, the most far-ranging oceanographic expeditions have as yet failed to bring back a single coil.

Of all the sea monsters whose legends have agitated the smoky air of waterside taverns, only the man-eating sharks have lived up to their reputation. There are some thirty species of man-eaters in the shark family, and they are popularly thought of as voracious, bloodthirsty, implacable, treacherous killers. Science, though it eschews such terms as "implacable" and "treacherous," considering them applicable to men not animals, has turned up a firm, factual foundation for every attribute.

Until a generation or so ago not very much was known about sharks. At one time, indeed, some reputable scientists even declared that sharks did not attack human beings. This notion was exploded when fishermen reported hauling in sharks with human remains in their stomachs, as occurred in Raritan Bay, New Jersey, in 1916. During the preceding two weeks, the fish — an $8\frac{1}{2}$-foot white shark — had apparently attacked no less than five individuals along the Jersey coast, four of whom died. At any rate, the attacks stopped after the animal was caught.

Even now, it is uncertain just how many of the three-hundred-odd species of sharks are man-eaters — or man-attackers. Many are too small to pose any threat to man (some are less than a foot long at maturity), and the very largest species are equally harmless. The whale shark, whose length of sixty feet or more makes it the largest living fish (whales run larger, but are, of course, not fish but mammals), is a toothless plankton-feeder, as is its somewhat smaller relative, the basking shark. Between the extremes is a whole range of sharks, from four to forty feet in length; some probably dangerous to man, others certainly so.

Many species of shark can be identified only by very close observation—and during a shark attack both the victim (assuming he survives to testify) and any on-lookers are likely to have other things on their minds. (It is likely, too, that chambers of commerce in seaside resorts discourage publicity, and therefore scientific investigation, about shark attacks.) The Greenland shark is not on record as having attacked a human being, yet it is known to eat seals, and one was found with the car-cass of an entire reindeer (albeit without horns) in its stomach. Perhaps the Greenland shark's reputation is based less on a pacific temperament than on the unlike-lihood of any man encountering it in the polar waters it frequents. With intensified study the list of dangerous sharks has expanded from a dozen or so to more than thirty—and will probably continue to grow. Meanwhile, the best advice to a swimmer or skin diver confronted by a shark is to get away first and worry about identifying the species later.

Scientific interest in sharks increased suddenly during World War II. With thousands of ship-sinkings and plane-ditchings in the tropical waters where sharks abound, the possibility of being chewed up added a peculiarly horrifying hazard, both physical and psycho-logical, to the normal perils of war. As not infrequently happens, some armed-forces spokesmen tried to cope with the problem by pretending it didn't exist; one national magazine was inspired to run an article entitled "The Shark Is a Sissy!" The sharks, however, failed to read it.

Other, more realistic, military leaders instituted a crash program to develop some means of protection against shark attacks. Scientists first began to review the little that was then known about sharks. It was well known that the fish have a curious combination of primitive and advanced biological characteristics. Their skeleton is composed chiefly of rubbery cartilage instead of true bone, and their spinal column extends into one lobe of their tail—traits that most fish lost at least 200 million years ago. Their reproductive system, on the other hand, is remarkably sophisticated. Fertilization is internal, as with reptiles and mammals. Instead of ejecting his sperm over the female's eggs, as nearly all fish do, the male shark injects it into the female's body, with the aid of a pair of organs misnamed "claspers." Nor are the fertile eggs simply cast upon the waters, with one chance in a million of survival. Some sharks lay eggs with tough shells (like those of reptiles) in which the larval shark can develop into a miniature adult in relative safety; in other species the eggs remain within the mother's body until hatching; in still others, the infants are actually nourished by the mother through structures resembling the mammalian placenta and umbilical cord, emerging only when they are fairly well able to fend for themselves.

All this, while interesting, was not very relevant to the problems of why and when sharks attack men and how they could be prevented from doing so. And here, unfortunately, the scientists collected a mass of folklore and grisly anecdotes, but few facts.

The war would not wait, however, so the investigators pushed ahead as best they could. At first, the best was not very good. Sharks were not affected by the chemical rotenone, known as a potent poison for most fish. Loud sounds failed to frighten the animals away; indeed, sounds resembling those produced by underwater explosions actually attracted sharks—a grim thought for survivors of torpedoed vessels.

A bit of lore from commercial shark fishermen provided the first promising lead. They had found that the rotting carcass of a dead shark seemed to drive away other sharks, though the animals were known to feed avidly on other types of carrion. And sure enough, when chunks of decayed shark flesh were offered to captive dogfish (a smallish shark harmless to man, though sometimes destructive to commercial fish), they not only refused to eat them but seemed to avoid even swimming near the bait.

Since sharks were known to possess an acute sense of smell—they are sometimes called "swimming noses"—it seemed probable that they were driven off by the odor of rotting shark flesh. The next step was to isolate the odorous substance that repelled the dogfish, since seamen and airmen could hardly carry chunks of ripe shark about with them in case of accidents.

At a site on the Florida coast where commercial shark fishermen were active, the scientists set up huge vats in which shark carcasses, covered with sea water, were left to rot. When the repulsive mass was analyzed chemically, a number of compounds were identified, including ammonium acetate. When dissolved in sea water, ammonium acetate releases acetic acid, the active ingredient in common household vinegar. It was then found that the dogfish shunned beef tainted with acetic acid in precisely the way they had avoided decayed shark meat. Another reasonably good repellent—this one not derived from sharks—turned out to be copper sulfate, which had long been known to inhibit certain kinds of marine animals. A combination of the two, in the compound copper acetate, appeared to be the most effective dogfish repellent of all.

But the chemical had yet to be tested under field conditions—not on dogfish, but on man-eaters. After an intensive hunt for good shark waters, involving, among others, Ernest Hemingway (who later dramatized an encounter between a shark and a fisherman in his famous novel *The Old Man and the Sea*), the scientists settled on the Gulf of Guayaquil on the coast of Ecuador. There a series of mangrove-bordered lagoons provides shelter to shoals of mullet and other fish, which in turn provide sustenance for numbers of sharks. Lines baited both with fresh mullet and porous bags of copper acetate were hung in the water—and the sharks avoided them, although they readily snatched at unprotected mullet bait.

At this point, the story becomes confused. According to one account, nigrosine, a soluble black dye, was added to the copper acetate as an additional antishark component, to disorient or discourage attacking sharks much as the octopus's ink cloud discourages smaller predatory fish. Another source has it that the nigrosine was added simply to give castaways visual assurance that the active ingredient—copper acetate—was in fact diffusing into the water.

Whatever the facts, the final product, christened "Shark Chaser," included both the acetate and the dye. Curiously, there seems to be no clear information on precisely how successful it was in repelling sharks. At the very least, however, it was a valuable morale booster for military personnel operating in shark-infested waters.

With the war's end sharks ceased to be of practical importance to most people. Scientific interest in the subject waned until the invention of scuba apparatus by the French naval officer Jacques Cousteau opened the sea's upper layers to exploration by marine biologists, archaeologists, geologists, and sportsmen. Inevitably, skin divers began encountering sharks—and their reports on the efficacy of Shark Chaser were, to say the least, disquieting. In 1951 Cousteau himself, diving with an associate in the tropical Atlantic, narrowly escaped three determined sharks. The two men vainly tried all the old sailors' recipes—gesticulation, shouting, releasing bubbles of air. Nor was the Shark Chaser any more effective: the most aggressive of the sharks "swam through the copper-stained water without a wink. . . . He seemed to know what he wanted, and he was in no hurry." Cousteau temporarily discouraged the beast, which was swimming directly at him, by clubbing it on the snout with an underwater camera. Fortunately, after a few minutes—which must have felt like a few hours—

the two men were able to make their way to the safety of the ship's launch. But their confidence in the shark repellent had been left behind in the water.

By 1958 enough similar reports had come in to make it clear that scientific knowledge of sharks was still in a pretty primitive state. At the urging of the U.S. Office of Naval Research, the American Institute of Biological Sciences set up a Shark Research Panel. Much of the panel's work was carried out at the Lerner Marine Laboratory in the Bahamas, where captured sharks could be kept alive in large salt-water pools. When necessary, they could be anesthetized by spraying their gills with a chemical, hauled out of water, and examined with no danger to the experimenter—or the shark.

Some of the studies merely confirmed that the shark's reputation as an acute smeller was well deserved. The animals were able to detect tuna juice (a predictably potent shark attractant) as much as seventy-five feet away, at which point the scientists estimated it had been diluted to a concentrate of less than one part in a million. By contrast, sharks whose nostrils had been closed with plugs containing a dilute anesthetic showed no interest in normally attractive substances.

On the other hand, the notion that the shark's vision is poor, and of little importance in feeding, turned out to be false, on both anatomic and behavioral grounds.

In its general structure the shark's eye is not unlike our own, with iris, adjustable lens, and light-sensitive retina. Originally, the retina was thought to contain no cone cells, which give the human eye both its sharp visual acuity (pattern discrimination) and its ability to perceive colors. More recently, conelike cells have been found in the shark's retina, though whether they enable the animal to see colors is still uncertain. Sharks certainly distinguish between patterns with great difficulty; Dr. Eugenie Clark, of the Cape Haze Marine Laboratory, was able to train lemon sharks to tell a square from a diamond, but not a square from a circle.

Other experiments, however, indicate that sharks easily learn to respond to differences in brightness because their retinas are well endowed with highly sensitive rod cells. Moreover, the shark's ability to function in poor light is enhanced by a remarkable layer of microscopic plates, lying behind the retina, called the *tapetum lucidum*. The plates, silvered with crystals of the compound guanine, act as mirrors. They reflect back through the retina any light that fails to react with the retinal cells, thereby giving it a "second chance" to make its impression upon the eye. It is probably the *tapetum lucidum* that enables the shark to feed at night, in turbid waters, and at considerable depths.

Some species that also feed during the day possess a further refinement of the tapetum: a layer of pigment cells interspersed between the mirrored plates, which, by expanding, cover the plates when the light is bright and, by contracting, uncover them in dim conditions. The change in the reflectivity of the retina can be clearly seen by looking into the eye of an anesthetized shark with an ophthalmoscope (the clicking instrument resembling a flashlight that is used by a physician to examine the human eye).

There is no longer any doubt that the shark relies

heavily on vision to find its prey. The Lerner investigators found that sharks temporarily blinded with plastic eyeshields had great difficulty in locating even the most fragrant morsel of food.

In hunting, the shark apparently also uses another sense, one somewhere between hearing and touch. It originates in the so-called lateral line, a sensory organ common to nearly all fish, but better developed in sharks than in many other species. It consists of slender tubes lying just beneath the shark's skin along both sides of its body and forming a network over the animal's "face." Connected to the surrounding water by smaller tubules, each tube contains clusters of sensory cells. These respond to vibrations in the water, apparently those of rather low frequency, such as might be caused by a large, injured fish — or a man, or a distant torpedo blast. One experimenter found, for example, that when a laboratory rat was placed in a shark tank, the fish showed little or no interest in the tiny splashes set up by the rat's swimming. But if the size of the disturbance was increased, by splashing the rat up and down on the end of a cord, the sharks would move in and often seize it.

The Lerner researchers believe that a shark is first attracted to a potential meal by vibrations perceived through its lateral line. As it moves toward the disturbance, it picks up the odor of blood or other bodily fluids, which it follows until the prey comes in sight. Then it switches over to vision for the final attack. Perry W. Gilbert, a member of the Lerner group, has eloquently described the feeding behavior of captive sharks:

When a large dead fish such as a 400-pound blue marlin is offered, lemon sharks first slowly circle it at a distance of six to 10 feet. Then, as they swim faster, the circle tightens and presently one shark moves in for the first bite. Contrary to popular belief, the shark seldom rolls on its side. Braking its forward motion with its large pectoral fins, the shark points upward slightly as its mouth makes contact with the bait. It opens its jaws wide, the lower jaw dropping downward and the upper jaw protruding markedly from beneath the thin upper lip. . . . Then it closes its jaws and shakes the entire forward part of its body from side to side until it has torn 10 to 15 pounds of tissue from the marlin.

Gilbert goes on to describe one of the most terrifying aspects of a shark's behavior, the so-called feeding frenzy. "As the blood and body juices of the marlin flow from the wound, the other sharks in the pack become more and more agitated and move in rapidly for their share of the meal. A wild scene . . . now ensues. The behavior of the animals appears to be determined entirely by the visual sense. An observer can substitute tin cans and wooden boxes for the marlin and the sharks will indiscriminately attack and consume them."

It is probably to the feeding frenzy that sharks owe their well-merited reputation for voracity, and in a certain sense, bloodthirstiness. The list of substances found inside captured sharks is long and improbable. Gilbert cites "a keg of nails, the skin of a buffalo . . . and a bottle of Madeira wine"; another author reports catching a shark that had swallowed, among other things, a three-foot-long roll of roofing paper with thirty feet of paper still wound on it.

Scientists have learned a great deal about sharks during the past decade, but thus far, they have failed to develop a wholly reliable shark repellent. They have discovered, for example, that copper acetate repels only the lemon shark and a few other species. A "bubble curtain" formed by releasing air through a submerged perforated pipe, once touted as a way of keeping sharks away from beaches, has proved even less useful; out of eleven tiger sharks tested, all but one swam through the curtain with no hesitation. Another possibility under investigation is a low-powered electrical device that would generate a "repellent" electrical field. One such apparatus worked well against lemon sharks, but actually attracted the more dangerous tiger sharks.

On the other hand, the nigrosine dye used in the original Shark Chaser seems to be quite useful even without the addition of ammonium acetate. As Gilbert describes one experiment, an open bottle of dye was placed in a circular tank with a free-swimming shark. The chemical gradually diffused through the water, coloring it black. "As the dye spreads from the center, the shark alters its pattern of swimming to avoid the dark area, restricting its movement to the spots that remain clear. By the time the dye bottle is completely empty, the shark has been penned in a small, crescent-shaped segment of the tank." Even so, the returns are not all in yet. A field test of an "improved" dye was abruptly abandoned when a shark not only swam through the murky cloud but bit off the end of the hose through which the repellent was being pumped into the water.

Gilbert and the other experts would be the first to agree that there is still a lot to learn about their predatory subjects. Some scientists, for example, believe that hunger as we know it has little to do with a shark's feeding; if the animals are fished with a baited line, the catches are highest when the animals have *full* bellies. On the other hand, some of the Lerner sharks, after consuming a large meal, showed no interest in baits that were highly attractive under other conditions. This seems only reasonable; despite the shark's legendary appetite and apparently ironbound digestion, some biological signal must tell him when he is chock-full.

A complicating factor is that many sharks seem to simply stop feeding for relatively long periods, living on the fat in their enormous livers. Some are thought to do so during the mating season (as do certain other fish), but others appear to stop feeding for no reason that has yet been discovered. It is no wonder that some experts observe sourly that the only *really* predictable thing about sharks is that they are unpredictable—or, to misapply a human term, "treacherous."

Until an effective shark repellent is found, experts can offer only negative advice to swimmers and skin divers in shark-infested waters.*

First, don't swim at night or in turbid water, where the danger of attack seems to be greatest.

Don't swim with a bleeding cut or other open wound; blood attracts sharks. For the same reason, spearfishermen should boat their catches immediately.

Don't—anywhere, any time—swim where dead fish or fish wastes are being thrown into the water—for example, near a fishing boat that is cleaning or sorting its catch. These substances not only attract sharks but send

---

* Almost no shark attacks have been recorded in waters colder than 70 degrees.

them into the feeding frenzy, from which escape is almost impossible.

If you see a shark nearby, don't get panicky and thrash around; this will only attract the fish. Swim toward shore as smoothly—and as rapidly—as possible.

If a shark gets close, don't strike it with your hand; its skin has the texture of coarse sandpaper (and was once used for the same purposes).

Don't provoke or attack a shark, with a spear gun, for example. The fish are incredibly durable and will almost certainly kill you before you can kill them. (The annals of shark fishermen are full of stories about sharks that, an hour out of the water or even gutted, still retained enough life to mangle a careless sailor.)

If sharks are in the neighborhood, don't assume you are safe in shallow water; sharks have mutilated people in water only a foot deep.

Finally, don't get neurotic about the "shark menace." Dangerous and terrifying though they are, sharks cannot in any sense be ranked as a serious threat to the average bather. On any summer weekend more Americans are killed by drowning than have died of shark bite during the past fifty years.

Sharks account for the great majority of dangerous animals that man is likely to meet in the sea. One of the others is the barracuda, which is not a man-eater, but has been known to deliver a very nasty bite to swimmers in its tropical habitat; in a very few cases the bite happened to strike a major artery, and the victims died from loss of blood. A handful of other fish are endowed with envenomed spines that can cause considerable pain, and in the case of a few tropical species such as the stonefish and lionfish, even death.

The largest group of unpleasant, though rarely dangerous, marine animals is also one of the most primitive. The coelenterates include the sea anemones, corals, jellyfishes, and the Portuguese man-of-war (which, despite its appearance, is not a jellyfish). Next to the sponges, they are perhaps the most simply structured multicelled animals on earth. All of them have stinging tentacles that they use to discourage predators, paralyze a prospective meal, or both. Their effective offensive and defensive apparatus doubtless helps explain why so primitive a group has survived so long (at least 600 million years) and so successfully (there are more species of coelenterates than there are of birds).

Many coelenterates are too small to disturb human beings, and almost none of them can cause more than an acutely unpleasant but transient skin irritation. Among the few exceptions is the Portuguese man-of-war. Whether the poison in this creature's tentacles can actually kill a man is uncertain, but it can certainly paralyze him—and death from drowning is quite as permanent as death from poison. Even more deadly, perhaps, is *Cyanea,* the "Lion's Mane" jellyfish, the effects of whose poison are accurately described in the Sherlock Holmes adventure of the same name.

Ironically, the most dangerous marine organisms next to the sharks are probably the microscopic dinoflagellates. This curious group of one-celled organisms, lying halfway between the plant and animal kingdoms, includes several species that under certain circumstances can kill more quietly but just as surely as a shark. The periodic explosive multiplication of these species has destroyed fish by the millions and is indirectly responsible for the illness of tens of thousands of Americans — of whom dozens have died.

A dinoflagellate bloom can be a dramatic spectacle, as witnessed by the account of one outbreak along the west coast of Florida. At first, the ocean turned a turbid yellow. Within a few months "the microorganisms had multiplied so abundantly that the water became thick and viscous. Turtles, barnacles, oysters, shrimp and crabs were killed. Stinking windrows of fish piled up for 60 miles along the beaches. Spray from the surf was so irritating to human beings that schools and hotels near the shore had to close."

When dinoflagellates become more than irritating to man, it is, oddly enough, because of the marine organisms they do *not* kill. Many shellfish, for example, can consume toxic dinoflagellates with no ill effects because they store the poison in one of their organs. But a human being who consumes a contaminated shellfish is not so lucky. One victim "awoke quite conscious but helplessly paralyzed. . . . She could not move in her bed or stand when lifted to the floor. She was scarcely able to speak for numbness about the face and mouth; was violently sick but unable to vomit; had a severe headache and backache and extreme dizziness." This victim recovered; occasionally, however, the poison paralyzes the breathing apparatus, producing death from suffocation.

Over one five-year period dinoflagellate poisoning

affected some forty thousand people on the Pacific Coast. Such incidents have led to considerable research on the "poisonous tides" of dinoflagellates, but thus far, the practical results have been meager. Dinoflagellates, like other marine microorganisms, proliferate because of a high concentration of nutrients in the water. A survey by Seymour H. Hutner and John J. A. McLaughlin established that nearly all reported dinoflagellate outbreaks have occurred where nutrients are brought to the surface by upwelling (as along the Pacific Coast) or by tidal turbulence in coastal regions (as in the Bay of Fundy). In other areas, such as the Florida coast, the outbreak seemed to be triggered by heavy rains on land, which were thought to have washed down nutrients from sewage and other sources.

But nobody has yet determined why a poisonous dinoflagellate species happens to multiply at a particular time and place, and a nonpoisonous species at another. The most the two researchers can say, after extensive laboratory experiment, is that "each species . . . requires just the right combination of temperature, salinity and light." And, as mentioned earlier, they are equally sensitive to minute differences in the concentration of particular elements (such as copper or cobalt) in the water.

Once a poisonous bloom has begun, however, it may grow by a sort of biological chain reaction. The poison kills fish, decaying fish increase the supply of nutrients, which means more dinoflagellates, which kill more fish, and so on.

Nobody even knows why some species of dinoflagellates are poisonous. The toxin does not appear to confer any special biological advantage, for several pairs of almost identical species have been found in which one is poisonous and the other harmless.

What is quite certain is that poisonous tides have been going on for a long time. Perhaps the earliest historical reference occurs in the Book of Exodus, which describes an outbreak in the Nile—most likely in the brackish waters near its mouth: "And all the waters that were in the river turned to blood, and the fish that was in the river died; and the river stank, and the Egyptians could not drink of the water." Evidence of a much earlier outbreak turned up a few years ago when McLaughlin met an oil geologist named W. R. Evett.

Like all oil geologists, Evett is an expert on fossil marine microorganisms. Their tiny skeletons occur by the trillions in rock deposits where oil is found, and the particular species discovered in a given layer of rock are routinely used to determine its geologic age.

In rock "cores" millions of years old, brought up from drillings in Pakistan and the Near East, Evett had found some unusual fossil dinoflagellates, a type of organism about which he knew little. He consulted McLaughlin to determine what they might be, and the biologist was able to show him similar living species from his own laboratory. It then emerged that the same rock layers also contained tremendous accumulations of fossil fish bones. Could dinoflagellates, asked Evett, kill fish? McLaughlin told the geologist of his and Hutner's researches; and when the dinoflagellate fossils were examined more closely, they seemed, as nearly as one could tell, to be of species thought to be responsible for fish kills in the recent past.

When a biologist finds a new poison, one of his first impulses is to ask what can be done with it. Not a few poisons have turned out to be valuable as drugs — curare, for example, has moved from the poison pots of South American witch doctors to hospital operating rooms, where it is used to relax a patient's muscles. Even if a poison has no medical use, it has often proved to be a valuable tool for the physiologist, becoming (in the words of one great physiological pioneer) "an instrument that dissociates and analyzes the most delicate phenomena of the living machine."

Dinoflagellate toxin has turned out to be just such an instrument. As the paralytic symptoms it produces would suggest, it blocks the nervous system, a property that it has in common with curare and with botulinus toxin (the lethal product of another microorganism that produces "ptomaine" poisoning). And the particular way in which it attacks the nerves, though meaningful only to an expert, is apparently just what the neurophysiologist needs for his laboratory. At any rate, this discovery has stimulated research that promises to expand our knowledge of how the body's control mechanisms work.

A sizable number of other marine organisms yield compounds of interest to the physiologist or the medical experimenter. Holothurin, a toxin secreted by sea cucumbers, contains a substance with physiological effects akin to those of the valuable heart stimulant digitalis; a modified version of the deadly stonefish's venom has been used experimentally to reduce blood pressure; some sponges manufacture antibiotics.

Shark blood, now under investigation at the University of Miami, might turn out to be the most valuable marine drug of all. Dr. M. Michael Sigel, noting that sharks rarely get cancer and are apparently resistant to most viruses, found that shark blood was useful in protecting chicks from the effects of a cancer-producing virus. When the blood was added to virus inoculations, tumors developed in only about one-third of the chicks, as against 90 per cent when the virus alone was injected.

If would be ironic indeed if mankind's greatest marine enemy should ultimately provide a remedy for one of mankind's most terrible and most recalcitrant diseases.

Does the shark have a rival as a marine menace in the legendary sea serpent? Perhaps the greatest authority on this controversial subject is the Belgian zoologist Bernard Heuvelmans. His book *In the Wake of the Sea Serpents* carefully evaluates all reports of large, unknown marine animals from Homer's day to the present.

The scientific fanatic, like the political fanatic, has certain easily detectable psychological traits. Notably, he insists on torturing even the most dubious or contradictory bit of evidence into support for his point of view. Heuvelmans shows none of these characteristics; his scientific sobriety makes him a persuasive witness. He sets aside reports that almost certainly refer to the giant squid, an animal once held in almost as low scientific repute as the sea serpent, although its existence is now accepted by the most conservative zoologists. He is left with 587 "sea serpent" sightings between the early seventeenth century and October, 1966. Of these, he disqualifies nearly half, either as "vague," "ambiguous," or otherwise suspect, or as outright false. (The field of what might be called sea herpetology has for more than a century suffered from the activities of practical jokers and waggish newspapermen.)

Heuvelmans freely admits that the three-hundred-odd remaining reports are contradictory among themselves, but only if one assumes that they were all inspired by a single kind of sea serpent. Closely examined, the reports fall into groups, each of which he believes shows a definite inner consistency. He suggests that they refer to no less than seven quite different possible "sea serpents"—none of them true snakes.*

Heuvelmans believes that two of his "serpents" are gigantic pinnipeds—relatives of the seals and the walrus; both have long, serpentine necks, and one has enormous eyes (for spotting its prey at moderate depths?) and a sort of mane, for which reason it is sometimes called the Merhorse. Three others, he suggests, are representatives, quite different in form, of a supposedly extinct group of whales called the Archeoceti. (The ancestry of the whales is one of the puzzles of evolution. The Archeoceti may or may not be the ancestors of today's whales, but judging from their skeletons, they were as fully adapted to marine life as are their present cetacean relatives. No fossils have yet been found of transitional forms between the whales and the land mammals from which they evolved. At least one of Heuvelmans' sup-

---

* Sea snakes of several species exist—but only of moderate size. They are poisonous, like some of their terrestrial relatives, but are considered unaggressive toward man.

posed Archeoceti, he thinks, represents just such a zoological "halfway house.")

He believes the sixth serpent to be an enormous marine reptile, resembling a lizard or crocodile. This conclusion is based not only on reports of the animal's shape but on the fact that, unlike the other "serpents," it has been sighted only in tropical waters — as would be expected if it were indeed a reptile rather than a warm-blooded mammal. One sizable species of modern crocodile is known to swim from island to island in the East Indies, and paleontologists have dug up fossil crocodiles fifty feet long.

Heuvelmans' seventh serpent is the most serpentine of the lot, having neither fins, mane, nor any other distinguishable appendages. It is, he believes, a "super-eel," or more probably, several species of eel, which differ among themselves as much as the everyday moray, conger, and common eel do.

Persuasive though they are, the Belgian's arguments have not been accepted by most zoologists. His critics, however, have been hampered by one basic problem: while it would be possible, in theory, to prove that sea serpents do exist, it is logically impossible to prove that they do not. Yet, with all respect to the difficulties facing the "antiserpent" faction, some of their arguments seem remarkably weak. One distinguished marine biologist rejects the hypothetical existence of a giant marine reptile by declaring that these creatures have been extinct for more than sixty million years. This, of course, is begging the question; the coelacanth had been "extinct" for just as long, until an African fisherman caught one.

No specimen of any of Heuvelmans' "serpents" has ever been landed, but one bit of evidence, preserved in a Copenhagen laboratory, is hard for even the most skeptical to explain away. In 1930 a Danish oceanographic expedition in the South Atlantic caught an eel larva not two or three inches long, like other eel larvae, but five feet long.

Now if the proportionate size of larva and adult is the same for this eel as it is for the common eel, a full-grown specimen of the Danish eel would be about ninety feet long and perhaps five feet thick. Even on the most conservative assumptions, the adult could not be less than fifteen feet long. Heuvelmans himself estimates, on quite different grounds, that his super-eel is perhaps fifty feet long, thus splitting the difference.

Someday, perhaps, somebody will haul up an adult super-eel for scientific examination. One expedition, seeking to do just that, actually caught something at twelve hundred feet that was big enough to bend a three-foot hook—and get away. Another sizable moving "something" has been spotted by sonar at an even greater depth, but there is no clue as to what it was. Until some other fisherman is more successful, the sea serpent will remain merely another intriguing fish story.

CHAPTER VI

BACK TO THE SEA

Having evolved in the sea, animal life took several billion years to get out of it. This long-delayed emergence is not surprising; the shift from marine to terrestrial existence poses the severest kinds of biological problems, requiring radical redesign of nearly all the body's organ systems. The more surprising fact is that of the groups that successfully redesigned themselves to meet the challenge of the land, most have later sent representatives back to the sea—which have met its quite different challenges with equal success.

The most numerous and diversified group of land animals—the insects—has profited very little from the sea's opportunities. Of the hundreds of thousands of insect species, only a tiny handful make their living at sea, and those, only around the edges. Even less maritime are the other terrestrial arthropods—scorpions, spiders, centipedes, and the like. The other major group of land animals, the vertebrates, are another matter. Except for the amphibians, which have never wholly adjusted to land life, and so far as we know have never been able to tolerate salt water, every group of terrestrial vertebrates—the reptiles, the mammals, even the birds—has sent species onto and into the sea. These groups of returning exiles have cropped up in evolutionary history not once or twice but at least a dozen times. Moreover, once back in the ocean, most of them have done so well that for the better part of the past 200 million years it is they (rather than such numerous but wholly aquatic groups as the fishes and squids) who have, at least in terms of physical size, ruled the waves, precisely as they have ruled the plains and mountains.

Every land animal that went back to the sea carried with it at least two major built-in handicaps. None of them could extract oxygen from the water, so they were compelled to seek it in the atmosphere at frequent intervals. And all of them, being habituated to fresh drinking water, must at the beginning have found salt water not merely unpalatable but harmful to their internal economies. To overcome these handicaps, many of the "returnees" evolved special and often intricate mechanisms for lessening their dependence on air and eliminating their need for fresh water.

The success of so many former land animals in making their way at sea against the competition of wholly marine groups suggests that life on land must have given them certain biological advantages that they could carry with them into the ocean. But in the case of one major group, the marine reptiles, it is hard to be sure precisely what those advantages were. Even more puzzling is the decline and extinction of this group after millions of years of successful maritime existence. With the other major group of returnees, the marine mammals, we are on firmer ground; the reasons for their success are reasonably clear, and the reasons for their decline and near extinction are even clearer—embarrassingly so, since our own species is responsible.

The return to the sea began soon after the appearance of the first wholly terrestrial vertebrates. Among the early groups of reptiles were the mesosaurs, looking something like mini-crocodiles, which pursued fish in lagoons and estuaries long before the dinosaurs walked the earth. But the heyday of the marine reptiles was the Mesozoic era, whose other name—the Age of Reptiles— applies to the sea quite as much as it does to the land. During this period at least six reptilian orders took up a marine life, in the process acquiring shapes both expectable and improbable.

Most expectable were the ichthyosaurs, whose name means fish-lizard (although they were not true lizards). In a remarkable example of what is called convergent evolution, they managed to evolve a shape very similar to the streamlined form that the fishes had evolved before them—and that another group of returnees, the cetaceans, would evolve independently much later. The head and body of an ichthyosaur looked much like those of a modern porpoise, except that the "beak" was considerably longer and there were four flippers rather than two. The tail, on the other hand, more resembled that of a shark, but upside down, for the ichthyosaur's vertebral column extended into the tail's lower, rather than its upper, lobe. Some of these predators were as long as fifty feet.

Considerably more *outré* were the plesiosaurs, which looked, as one paleontologist has put it, rather like a snake threaded through a turtle. Hitched to a stubby, arched body with four turtlelike flippers was a relatively slender tail and a long—sometimes very long—neck, with a small head. As swimmers, they were hardly in the ichthyosaurs' class, but their flexible necks must have given them a "reach" and maneuverability sufficient to keep their bulging bellies filled with fish.

The third major group of Mesozoic marine reptiles were true lizards, though gigantic ones. The mosasaurs, in fact, rather resembled enormous monitor lizards, but with broad, flattened tails to propel them through the water.

The list of marine reptiles does not stop there. Placodus, probably a distant relative of the plesiosaurs, but lacking their typical serpentine neck, had blunt, peglike teeth, suggesting that it lived, like the modern walrus, by grubbing along shallow bottoms for mollusks and similar fare. The crocodiles also made their appearance; some of them, like certain of their modern descendants, seem to have lived in tidal estuaries and lagoons while one group, the thalattosuchians, was truly marine, having traded paws for paddles and evolved a long, ichthyosaurlike tail for propulsion. The turtles took to the sea, too; one of their marine representatives attained a length of better than twelve feet—twice the size of its largest modern relatives.

In making their way back to the sea, all these reptiles confronted the biological handicaps mentioned above—and one other. Reptiles typically lay eggs, and these will not hatch underwater, since the developing embryo must take in a certain amount of oxygen through the permeable shell. The ichthyosaurs, whose thoroughly fishlike forms indicate that they were open-sea animals, solved this problem by hatching their eggs inside their bodies and bringing forth their young alive. We know this not simply by analogy with living reptiles (some land snakes, and most sea snakes, bear living young) but from fossils of pregnant female ichthyosaurs. In one case, some of the young had just emerged when the family was overtaken by a prehistoric catastrophe.

The marine turtles undoubtedly went ashore to lay their eggs, as their modern descendants still do, and the same may well be true of most of the other groups of reptiles that returned to the sea. It is perhaps significant that the whole plesiosaur group, and its putative relative Placodus, possessed a bony covering on the abdomen, formed by the basketlike interweaving of extensions to the ribs; this was sometimes supplemented by enormous, platelike shoulder and pelvic bones. Doubtless this hard underbelly served to protect the animals against the assaults of other predators, but it may also have protected their internal organs from being crushed by the animal's weight when—and if—they ventured onto a beach to lay their eggs. A plesiosaur on land would certainly have moved very clumsily—but perhaps not much more so than a modern walrus or elephant seal.

The more subtle adaptations made by prehistoric reptiles to a marine life can only be guessed, since they involved the animals' soft tissues, which have left no fossil records. However, some possibilities emerge from a consideration of these adaptations among modern reptiles.

A notable fact about aquatic reptiles—both marine and fresh-water—is that they can remain underwater for much longer than any mammal. Marine turtles, when "resting," can stay submerged for hours at a time, and some fresh-water species can do without air for days or even weeks. In part this is because reptiles, which need not maintain a constant body temperature as mammals must, have less active metabolisms and so require less oxygen. But in some cases the need for air has been further reduced by more radical measures.

In certain sea snakes, for example, one lung has become so enlarged that it takes up a sizable portion of the animal's bodily cavity. This provides a reserve supply of oxygen—and may also provide buoyancy, as a fish's air bladder does. Even more remarkable are the underwater adaptations of some fresh-water turtles, including the pond slider and cooter of the southeastern United States, which are probably the world's champions at remaining submerged for long periods.

This capacity had long been ascribed to the peculiar construction of the animals' urinary tract, which includes two supplementary "lateral" bladders in addition to the normal urinary bladder. Noting that while underwater the animals take in and squirt out water through their cloaca (the vent through which reptiles excrete all their body wastes), scientists suggested that they absorbed oxygen through the walls of the lateral bladders, which thus served as sort of crude gills.

Eugene Robins, a physician and physiologist at the University of Pittsburgh, set about testing this theory. He and his associates procured a number of pond turtles and plugged up their cloacae with plaster. But the animals were still able to survive underwater for as long as two weeks. Clearly, whether or not the turtles *could* absorb oxygen through their cloacae, they didn't have to.

A second experiment established that they not only didn't have to but never did. When turtles with unplugged cloacae were immersed in water with an abnormally high oxygen content, the oxygen concentration in their blood did not rise, as it presumably would have done had they been obtaining oxygen from the water. Instead, blood oxygen rapidly fell to near zero, regardless of how much—or how little—oxygen the water contained.

In still another experiment, the Pittsburgh researchers injected the turtles with sodium cyanide. This poison almost completely blocks the oxidation of food compounds, such as sugars, within the body's cells—and thus is quickly fatal to most reptiles and all mammals. The turtles, however, were unaffected by normally mortal doses.

The turtles, evidently, were able to do without oxygen for long periods because they could somehow "switch over" their entire metabolism to a process requiring no oxygen. Other experiments made clear that this process was what is called anaerobic glycolysis, a common enough biochemical mechanism in nearly all higher animals and part of the normal way in which cells obtain energy by breaking down simple sugars—notably glucose. In the process, a molecule of glucose is split into two molecules of pyruvic acid, with the liberation of a certain amount of energy. Normally, however, the pyruvic acid is then oxidized, producing a great deal more energy.

There are two drawbacks to anaerobic glycolysis. First, it is inefficient, making available to the body not much more than one-twentieth of the potential chemical energy of the glucose molecule. But inefficient though it is, it evidently supplies enough energy for the sluggish metabolism of a turtle. The second drawback is that when pyruvic acid is *not* oxidized, it is transformed into lactic acid, which, like pyruvic acid itself, is a fairly strong acid — strong enough to disrupt even a turtle's physiology.

It had been known for a long time that the turtle's abdominal cavity contains a high volume of remarkably alkaline fluid, which has in it large quantities of bicarbonate — the perennial household remedy for acid stomach. Robins was able to show that this fluid neutralizes the lactic acid, thus acting "as a sort of built-in ampoule of bicarbonate of soda to hold down acidosis as and when it develops."

Unfortunately, we do not yet know whether marine turtles employ the elegantly evolved biochemical mechanisms of these fresh-water species to extend what might, in military jargon, be called their "submersion capability." And of course we do not, and cannot, know whether any of the extinct marine reptiles had evolved similar devices. Some of them quite possibly did; in others, the solution to the breathing problem may have been much simpler. The plesiosaur's long neck, for example, could well have served not only as a grab for seizing fish but also as a snorkel tube, by means of which the animal's head could be pushed above the water for a quick gulp of air while its body remained submerged.

In the case of the other big metabolic problem attached to a marine life—salty drinking water—the answers are a good deal less problematical. We *know* how at least two groups of modern marine reptiles manage this, and we can be almost certain that the extinct groups *must* have evolved some similar technique.

The drinking-water problem that confronts a land vertebrate in the sea is typified by the human castaway of fact and fiction. As most people know, a shipwrecked sailor drifting on a raft cannot slake his thirst with sea water; many people, however, do not know why.

The problem is with the salt—and with the kidneys, which eliminate excess salt from the body, as they do many other unneeded or potentially toxic substances. All these compounds are flushed away, dissolved in the urine. But because of the physiological limitations of the human kidney, the urine can contain, at a maximum, only 2 per cent salt. Sea water, however, contains more than 3 per cent salt. Thus the kidneys, seeking to rid the body of the excess, must draw on the body's own reserves of water to dilute the salt to the 2 per cent level. As a result, the castaway drinking sea water will very quickly find himself a great deal dryer, and thirstier, than he was before. (This same phenomenon explains why many bars serve their patrons salty foods like nuts and pretzels.)

In terms of kidney function, marine reptiles and sea birds (in some important physiological respects, birds are merely reptiles with feathers) are even worse off than man: their kidneys can handle a salt concentration of less than .5 per cent. Yet many species spend months, or even their entire lives, with no possible access to fresh water.

Some physiologists have held that these marine animals obtain the water they need from their diet of fish. But this really explains nothing. To be sure, most fish are less salty than sea water—to the point, in fact, where human castaways have been advised to supplement their drinking water by catching fish and squeezing the juice out of them. But while this would probably do well enough for a man, with his relatively efficient kidney, it fails to work for a gull or marine turtle.

Some years ago Knut Schmidt-Nielsen, a Danish-American physiologist specializing in animals with out-of-the-ordinary drinking habits (he has also studied the camel), administered large quantities of sea water to captive sea gulls—in one case, the equivalent of a two-gallon drink for a man. The birds showed no ill effects. Their urine output increased sharply, but, as expected, by no means eliminated all the salt. The excess showed up instead in a clear, colorless fluid that dripped from the birds' beaks. Chemical analysis proved this fluid to be an almost pure 5 per cent salt solution—quite concentrated enough to rid the bird of whatever salt it took in from drinking, or eating, at sea and leave a residue of desalted water to meet the animal's physiological needs.

The source of this salty drip turned out to be the so-called nasal glands found in the heads of all birds. Anatomists described these glands more than a century ago and noted that they were considerably larger in sea birds than in land birds. Some had speculated that the sea birds' glands secreted "tears" to wash salt water out of their eyes. In fact, as Schmidt-Nielsen discovered, the glands "wash" salt out of the sea birds' bodies.

Microscopic examination of the salt glands revealed that in some important structural features they are very similar to the mammalian kidney — the only organ that can even approach their capacity to concentrate salt. Apparently, however, they are much simpler organs than the kidney; where the latter performs a wide variety of regulatory and eliminative functions, the salt glands seem to exist only to remove salt.

Schmidt-Nielsen was able to show that salt glands were present in a dozen different species of sea bird, including all the major orders with marine representatives, though the glands' location varied somewhat, as did the means by which the salt was expelled. A particularly elaborate adaptation was found in the petrel, which, along with its relatives, the shearwater and the albatross, has tubular nostrils lying along the top of its beak. (The group is often called the "tubenoses.")

The nostrils serve as a sort of water pistol through which the petrel (and presumably the other tubenoses) forcibly ejects droplets of salt solution. It has evidently evolved this mechanism because, unlike other sea birds, it flies almost continuously during the months it spends at sea. The "pistol" permits it to eliminate salt in flight without having the solution blown into its eyes.

Having discovered how sea birds solve their drinking problem, Schmidt-Nielsen turned to modern marine

reptiles, which, as he knew, had much the same inefficient kidneys as birds. He noted that turtles sometimes
"weep" (indeed, even the Mock Turtle in *Alice in
Wonderland* wept, and almost continuously at that, because he was not a real turtle). Archie Carr, probably
the leading authority on marine turtles, has described a
female coming ashore to lay eggs and noted that she
"began secreting copious tears shortly after she left the
water, and these continued to flow after the nest was
dug. By the time she had begun to lay, her eyes were
closed and plastered over with tear-soaked sand, and
the effect was doleful in the extreme."

Regrettably, Carr succumbed to anthropomorphism
in this case, a weakness that sometimes overcomes even
the most able naturalists. There was nothing doleful
about the female turtle's plight. Nor (as his account made
clear) was she "weeping" to wash sand from her eyes,
as some earlier naturalists had suggested, or because
of the alleged "pain" of laying eggs. When Schmidt-
Nielsen analyzed the tears of a large marine turtle, he
found that they were chemically similar to the secre-
tions of the sea birds' salt glands and that they originated in an organ of much the same structure. He then
found salt glands in the heads of the marine iguana,
which browses on salty seaweeds along the shores of
the rather arid Galápagos Islands.

The location of the turtles' salt glands indicates that
the organs evolved in a different manner from those of
birds; the iguana's may represent still another evolutionary line. Thus it would appear that two, and perhaps
three, modern groups of animals have independently

evolved a special mechanism for getting rid of unwanted salt. There seems to be no doubt, then, that the great Mesozoic marine reptiles *could* have evolved similar devices — and, in my opinion at least, there is very little doubt that they *did.* Unless their kidneys were far more efficient than those of any modern reptile (and most modern mammals), some sort of salt gland seems to be the only answer.

That the marine reptiles adapted to the special conditions of life in the sea does not explain their great evolutionary successes. They may have been less dependent than land reptiles on continuous access to air, but they were still dependent on it, as their fish contemporaries were not. Their salt glands (or whatever) must have enabled the seagoing reptiles to do without fresh water, but any marine fish can do as well. Yet, all the fossil evidence indicates that it was the returning reptiles, not the fish or any other purely marine group, that dominated ocean life all during the Mesozoic era. The great sharks did not appear until later; the largest of these — a monster relative of today's white shark, which may have reached eighty feet — dates from only about twenty million years ago, when the great marine reptiles were long gone.

Conceivably, the reptiles' sensory equipment was better. As a group, they see considerably better than fish do, and the ichthyosaurs, at least, seem to have depended heavily on sight, for their eyes were the biggest, proportionate to their size, of any animal before or since. But sight is of but limited use in the sea, with a range under optimum conditions of only a hundred feet or so and of much less than that in turbid or darker waters. Reptiles could probably also hear better than fish (there is, in fact, some doubt whether fish can hear at all), but they had, in exchange, lost the lateral line that serves fish as an apparently effective substitute for hearing.

Again, the reptiles' control systems—which is to say their brains—may have been better than those of the fish around them. Present-day reptiles, at least, though far from brainy by mammalian standards, have larger and apparently better brains than fish do. With extinct reptiles, we cannot be sure. In a mammalian fossil the interior of the skull fairly accurately indicates the size and shape of the brain, but reptile skulls are constructed differently, so that the braininess—even in the literal sense—of the Mesozoic sea lords must remain a matter of speculation. Whatever the reasons for their success, however, the marine reptiles ruled the oceans for more than 100 million years; the mammals, by contrast have dominated life on earth for fewer than 70 million years.

Even less explicable than the marine reptiles' long dominance is their disappearance. The ichthyosaurs vanished even before the end of the Age of Reptiles, and by the time the subsequent Age of Mammals opened, the great marine reptiles had disappeared without a trace—except for the crocodiles, none of which are more than quasi-marine, and the turtles. Presumably they disappeared for the same reasons the dinosaurs disappeared, but nobody knows what *those* reasons were either. It is sometimes said that the Mesozoic reptiles were wiped out by competition with the mammals—but in fact, as the great paleontologist George

Gaylord Simpson has pointed out, mammals big enough to compete effectively with the giant reptiles did not appear until long after those "competitors" had already vanished.

It has also been suggested that the reptiles were wiped out by climatic changes, for at the very end of the Age of Reptiles there was a slight, albeit relatively temporary, cooling over most of the globe. But there is grave doubt as to whether this cooling was sufficient to seriously disturb the balance of life, even on land. Similar climatic shifts had occurred earlier in the Mesozoic, with no catastrophic effects on land animals. And in the sea, as we have seen, climatic changes are far less marked than on land.

The fact is that we shall quite possibly never know why *any* of the great reptiles disappeared.

With the reptiles gone, the mammals—who had been hanging about in the wings all during the dinosaur age—expanded and differentiated to fill the various biological roles that had been left vacant. Some of them took to the sea, as the reptiles had done before them, and often with equally marked success. First on the marine scene were the cetaceans, which evolved into the various varieties of whales, both large and small (*i.e.*, the porpoises and the dolphins). Somewhat later, the Carnivora sent a large branch to sea—the related seals, sea lions, and walruses and, as a separate branch, a single species of sea otter. A third, less conspicuous group was the Sirenia, which includes the manatees and the dugong.

In contrast to the reptiles, there is no great mystery about how the mammals managed to adapt to marine life.

The problem of where, or whether, to lay eggs did not concern them, since all mammals bring forth living young. Nor was fresh water a serious difficulty; their kidneys were reasonably efficient to begin with, and they had no trouble in stepping up that efficiency enough to cope with sea water and sea food.* The oxygen problem was less easily solved. Because of their constant body temperatures, and for other reasons, mammals require more oxygen than reptiles do, and no marine mammal can match the underwater endurance record of the sea turtle—let alone that of Dr. Robins' pond turtles. Still, a seal can remain submerged for some twenty minutes—about ten times as long as a man—and the larger whales can stay down for up to an hour and a half.

How the whales do it has never been determined; they are not good laboratory animals. The seal, however, has been extensively studied, and its oxygen-conserving mechanisms turn out to be quite as interesting, if not as effective, as those of the turtle.

It has been known for at least half a century that when seals dive, their heartbeat slows markedly, and some physiologists thought that the resulting reduction in blood circulation would conserve enough oxygen to account for their ability to stay underwater for prolonged periods. Over the years, however, doubters have pointed to the fact that the mammalian heart and brain are peculiarly sensitive to a lack of oxygen. A few seconds without it produces unconsciousness; a few minutes brings on severe and permanent brain damage—and

* Schmidt-Nielsen has found that several desert animals, who face a different kind of fresh-water problem, have independently evolved super-efficient kidneys. One of them, the kangaroo rat of the American Southwest, can survive quite happily in the laboratory on a diet of sea water and dried soybeans.

there was no reason to suppose that the seal's brain was any exception. Moreover, it was found that the seal's slower heartbeat did not produce a drop in blood pressure — as it should have done, other things being equal.

The only reasonable explanation seemed to be that the animal maintained its blood pressure by, in effect, cutting off circulation to certain parts of its body. With less tissue to pump blood through, the heart could maintain the pressure even while beating slower. This hypothesis was proved by the same Eugene Robins who had studied submersion in turtles. He employed a relatively new medical technique known as angiography, in which it is possible to study blood circulation by using x rays. Blood as such does not show up on an x-ray plate, since it does not "contrast" with the body's soft tissues as bones do. But by injecting into the blood a dye that is relatively opaque to x rays, the physician can obtain a clear picture of the blood flow in any part of the body.

Robins felt that angiography would provide a definite answer to what was happening in the seal's circulation when it dived. He conducted his experiments on captive harbor seals that were restrained in a sort of canvas strait jacket and then strapped to a "diving board" — actually, a small seesaw. When the seesaw was tipped forward, the animal's head was immersed in a tank of water. Robins noted that music appeared to have a soothing influence and that the seals seemed to show a strong preference for light classics, as against rock 'n' roll.

Angiograms made while the seals were submerged showed clearly that the arteries serving the liver, spleen, kidneys, flippers, and some other parts of the body quickly went into spasm, shutting off the blood flow to these areas. To the x ray's eye, indeed, it was as if these organs had been surgically removed. The brain (and possibly the heart) were unaffected.

It appears, then, that seals conserve oxygen during a dive by circulating it only to the most vital organs—the brain, and probably the heart. The skeletal (swimming) muscles—which the animal needs while diving under natural conditions—evidently derive energy from anaerobic glycolysis, as the diving turtle does. Other organs, such as the digestive tract, may employ the same device—or may go into temporary suspended animation.

The muscles of a human runner, which are using up more oxygen than the input of the lungs can supply, shift over partially to anaerobic glycolysis while also using such oxygen as is available from the blood that continues to flow through them. Lacking the seal's circulatory cutoff valves, man cannot conserve oxygen as the seal does—and therefore, if he shifts from running to diving, he cannot stay under nearly as long. The seal, on the other hand, cannot stay under as long as the turtle, in part because his brain cannot do without oxygen for as long a time as the turtle's can.

Their brain is the reason why the ancestors of today's marine mammals so successfully managed the transition to marine life. As a mechanism for controlling and guiding the body, the active—and oxygen-hungry—mammalian brain is far more efficient than that of the most advanced reptile, let alone, the fish, squid, or mollusks on which most marine mammals feed. The porpoise, whose physiology has been extensively studied, actually has a brain somewhat larger, in proportion to its size, than that of man. Whether it is as "good" as a man's brain is still uncertain. Intelligence tests devised for humans are for the most part unsuitable for porpoises, and reports that porpoises can "talk" in anything approaching the human sense appear to be premature, to say the least. But there is no doubt that the creature's intelligence is well above the mammalian average and that its brain has developed some remarkable special adaptations. These enable it to perform, unaided, feats of underwater "object location" that we can carry out only with elaborate electronic devices—and less efficiently at that. Indeed, the U.S. Navy is presently engaged in training porpoises to help recover objects lost underwater.

The porpoise has for centuries been deemed unusually friendly to man—a belief doubtless strengthened by the shape of its mouth, which gives it a built-in, permanent grin. The many legends about porpoises are typified by the Greek fable of Arion, who jumped overboard from a ship when the sailors threatened to murder him for his money, and was borne ashore on a dolphin's back. Unlike most animal fables, however, the classical porpoise stories almost certainly are based on fact. Whether the animal ever saved anyone's life is uncertain—although a Florida woman who fell off a boat near Grand Bahama Island in 1960 swears that a porpoise prevented her from drowning by nudging her toward shallow water. But porpoises (such as the famous "Opo" of New Zealand) are known to have played with human youngsters and on occasion given them rides. ("When I was standing in the water with my legs apart," one young woman recalled, "[Opo] would go between them and pick me up and carry me a short distance."

The average porpoise is rarely so actively involved with man, but all of them do seem to be what one authority calls "man-oriented." They seek the proximity of human beings in a manner that seems to reflect a friendly curiosity, and seldom if ever are they hostile or vicious to man. Though few of them play with children, all of them seem to enjoy playing on their own or among themselves; this puts them among the minority of mammal species (man is one of the others) that, *as adults*, engage in activities for their own sake, with no apparent practical end like feeding or reproduction.

Research into the engaging porpoise was intensified during and after World War II, when underwater sound equipment designed to detect submarines also picked up the sounds of fish, shrimps, and porpoises.

Porpoises generate two general types of noises. One is a sort of squeaky whistle or chirp by which they appear to communicate with their fellows. Skin divers have reported that female porpoises respond to the chirps of their own young amid the dissonant squeaks of a porpoise "nursery." Injured porpoises squeak to attract other porpoises, and healthy porpoises have been seen on occasion to swim on either side of a distressed animal, supporting it so that its blowhole will not sink below the surface of the water.

That porpoises communicate among themselves is not particularly remarkable, since many—perhaps most

—mammal species do so by various means. Much more remarkable is the animal's other set of sounds, variously described as sputtering, rasping, woodpeckerlike tapping, and "the rusty hinge sound." At times, these noises somewhat resemble the classic Bronx cheer.

The cetologist Winthrop W. Kellogg and his associates have made extensive studies of porpoises, picking up the sounds of captive animals with underwater microphones and analyzing them electronically. The "rasping" noise turned out to be a series of sharp clicks. Some were as slow as five per second, in which case each click could be distinguished as "a sharp staccato report—like that produced by striking a heavy wooden table with a small hammer." At other times the clicks came so rapidly that the individual sounds merged into a rasping, groaning, or even barking noise.

The Kellogg group established that porpoises use these clicks in a manner precisely analagous to electronic sonar—to locate and take the range of underwater objects. The porpoises in the experiment would typically begin signaling whenever the researchers produced a splash on the surface of the tank. "If a splash was made alone . . . the sputtering stopped after a few seconds. When a splash was followed by the presence of some new object in the water, exploratory sound signals continued—presumably until the size and distance of the object had been determined." The animals could apparently detect objects as small as a BB shot, about one-sixth of an inch in diameter. Even when there were no splashes, the animals would emit short bursts of signals every two seconds or so, which enabled them to detect objects placed in the tank without a splash. The researchers compared this behavior to a land animal periodically glancing about its surroundings.

The porpoises quickly learned to distinguish—at a distance too great for smelling—between a twelve-inch fish of a type they disliked and a six-inch species that they found palatable, showing no interest when the former was tossed into the water, but swimming quickly to the latter. They also—and invariably—showed no interest in an otherwise attractive fish when access to it was blocked off by an invisible sheet of plate glass. Their most remarkable feat was the negotiation of an "obstacle course"—an array of thirty-six metal poles, about eight feet apart, lowered into the tank and so arranged as to "ring" audibly if the animals touched them. The first couple of times, the porpoises flicked a few poles with their tails in passing—apparently because they were unused to the rather sharp turns needed to negotiate the labyrinth; thereafter, they swam through the forest of poles without grazing them, even on the darkest nights. Indeed, they swam considerably faster with the obstacles in place than they did at other times—conceivably because they were "playing" with them. Just possibly, the poles may have constituted the porpoise equivalent of a Jungle gym.

Other experiments have established that porpoises are not only highly adept at analyzing and interpreting the sonar signals they emit but can also solve certain problems much more quickly than the chimpanzee, hitherto considered the intelligence champion among subhuman animals. These findings have led to suggestions that porpoises might actually be trained to "herd" fish for human fishermen, assist sailors in distress, and perform other useful tasks at sea, as man's domestic animals have long done on land. This may sound fanciful—but in the light of what we now know about these remarkable and appealing creatures, it is not much more improbable than the many other fantasies of science fiction that have subsequently turned into scientific fact.

The marine mammals have faced, and solved, one other problem that the reptiles did not have: the need, as warm-blooded animals, to maintain their body temperature. As they evolved over millions of years, the earth's climate was growing steadily colder, and unless, like the marine turtles, the marine mammals were willing to restrict themselves to the shrinking belt of tropical and warm temperate seas, they had to evolve a means of insulating themselves against chilling, for water, as noted before, can absorb more heat than any other common substance.

The marine mammals solved the problem perfectly, for nearly all of them have long exploited the frigid but rich oceanic feeding grounds around Greenland, Alaska, and Antarctica. Ironically, the very means by which they have preserved their lives against the cold of polar waters have indirectly caused the deaths of millions of them and brought several species to the edge of extinction. The thick pelts of the fur seal and the sea otter, coveted by chilly ladies of fashion, led to such wholesale slaughter that both species were nearly wiped out before international control forced more rational harvesting of the fur crop. And the thick, insulating blubber of the great whales, serving first for lamp oil and later transformed into margarine and soap, has led whalers to deplete the population of these magnificent mammals literally to the vanishing point.

Unlike the decline of the marine reptiles, the reason for the decline of the marine mammals is no mystery. The history of man's assault upon the natural resources of his planet is, God knows, replete with horrible examples of greed and stupidity. But even among these dismal annals, the performance of the whaling industry has set some sort of all-time low.

The first great whaling boom took place in the eighteenth and nineteenth centuries. Not a few New England fortunes were founded on the rich cargoes of oil and whalebone (used for umbrellas and corsets) brought back to New Bedford, New London, and Nantucket. *Moby Dick*, which many consider the greatest American novel of the nineteenth century, has a whale in the title role—and is, incidentally, quite as distinguished a treatise on whaling as it is a work of the imagination.

Melville himself wondered whether man's furious pursuit of the whale would not destroy the species, but he convinced himself that it would not. Chapter 105 of *Moby Dick* is a brilliant pioneer effort in the rationalizations by which the whaling industry has perennially befuddled itself and the world. In fact, even the open-boat whaling of Melville's day soon managed to nearly wipe out both the bowhead whale and the right whale; what saved them from outright extinction was not the whalers but the replacement of whale oil by kerosene as a lamp fuel. Although they have not been caught systematically for nearly a hundred years, they are still very rare today.

Around the turn of the century, chemists discovered how to process the malodorous whale oil into edible fats and bland soap; at the same time, major technological improvements in whaling enabled whalers to catch species, such as the blue and finback, whose speed, or tendency to sink after being killed, had previously saved them from pursuit.

As early as the 1930's there were ominous signs that these species also were being hunted too intensively. The situation became really serious, however, after World War II. Japan and the U.S.S.R. began building new whaling fleets—and, as future developments showed, they were determined to make these fleets pay for themselves, down to the last yen or ruble.

At the same time, an International Whaling Commission was set up to prevent overexploitation of the whale fisheries. None of the whaling nations, however, were willing to give it either inspection or enforcement powers; moreover, they insisted that quotas be agreed on unanimously—meaning that the catches must be big enough to satisfy the greediest and most shortsighted. Worst of all, perhaps, the quotas were not assigned to particular species but to whales *as a whole*—meaning that a nation might take even the rarest species in filling its quota.

The first whale to feel the impact of these nonregulatory regulations was the mighty blue—the largest animal ever to evolve on the earth. Some blues have been known to reach 110 feet in length and a weight (when weighed in sections — nobody has ever figured out how to weigh an entire whale) of 150 tons. Blue-whale catches had been dropping since the early 1930's, from more than 20,000 during the Antarctic summer whaling season of 1932-33 to less than 2,500 in 1953-54. But the whaling nations were not alarmed by these figures — they simply stepped up their catches of the closely related finback whale.

In 1963, when the blue catch had dropped below 1,000, the commission at last acted. Under strong pressure from some of the nonwhaling nations, backed by the recommendations of an expert fact-finding committee, it forbade the hunting of blues anywhere — except in more than three million square miles of the best hunting grounds, which the Japanese insisted must be exempted. At the same time, the whaling nations rejected the committee's proposed limit on the total whale catch, subsequently setting their own quota at double the recommended figure.

While the blue-whale catch continued to fall, the finback catch also dropped, from nearly 28,000 in 1960-61 to less than 5,000 in 1965-66. (Ominously, too, an increasing proportion of the finbacks were smaller, immature animals.) The yield of Antarctic whaling became so modest that both the British and the Dutch ceased to send fleets there; more recently, the Norwegians have followed their example.

The Japanese and Russians, however, are still active; their wage levels are low enough to keep whaling profitable, and "democratic" Japanese capitalists and "authoritarian" Soviet commissars are united in ignoring biological, or even economic, logic. True, the blues are now—in theory—under total protection by the commission; one can hope (though not be certain) that the Japanese and Russians are observing the regulations in practice. But international regulations apply solely to open-sea whaling; only after considerable persuasion were Chile and Peru, whose citizens pursue whales from shore stations, induced to protect the blues—temporarily. The finbacks are only partially protected and the sperm whale—the one remaining fairly numerous species of large whale—is for all practical purposes not protected at all.

Even if not a single blue were caught from now on, there are probably too few left to ensure that, in the vast wastes of the ocean, male blues will meet females sufficiently often to maintain the species. The humpback whale (which, like the blue, was protected only when it was almost fished out) is on the brink of extinction, and at least two other species are close behind. The Russian and Japanese whalers are therefore concentrating their efforts on the sperm whale, and one needs no crystal ball to foresee the result. At least one American expert is convinced that they will play out the tragedy "to the last whale."

It took many millions of years for nature to wipe out the great marine reptiles; man is performing the same service for the great marine mammals in less than three centuries. Considered as a food resource, they are insignificant; it is unlikely that a single human being will miss a single meal when the large marine mammals join the mammoth, the dodo, and the passenger pigeon. Biologically, however, they are unique; the largest animals on earth, and judging from their small cousin the porpoise, probably among the most intelligent next to man. (Or is that qualification, in view of what I have written, really accurate?) Nobody knows what questions about animal behavior and animal intelligence the whales could answer for us; it now seems miserably probable that nobody ever will.

CHAPTER VII
THE CONTINUING SEARCH

Beneath the bended arm of Cape Cod, like the trailing sleeve of a 1910 evening gown, hang the Elizabeth Islands, the tops of a range of gravelly hills thrown up by the last ice sheet. The island chain is broken by passages—"holes"—of which the northernmost is Woods Hole, a twisting, boulder-strewn channel through which the seesawing currents between Buzzards Bay and Vineyard Sound rip at speeds of up to seven knots. Facing the hole, on the southernmost peninsula of the Cape itself, is the small community of Woods Hole, the most important center of oceanographic research on earth.

To get a picture of the extraordinary scope of current oceanographic research around the world, one can do no better than wander about Woods Hole, viewing its scientific exhibits, chatting with its scientists, and leafing through its publications. What follows is merely a sampling; a comprehensive report on the activities at even this single center would fill a large book. In keeping with the theme of this book, moreover, it is a sampling heavily weighted toward biology and ecology —but with enough about physical oceanography to suggest the fascinating variety of research taking place in that area.

Like many major scientific centers, Woods Hole was not planned but simply grew. However, its growth was not unconnected with its location within a few hundred miles of almost every imaginable kind of marine environment. A few hours north through the Cape Cod Canal are the shallow waters and tidal flats of Cape Cod Bay; to the east lie the shifting sandbanks around Nantucket. East again and then north, around the elbow of the Cape, are the rich fishing banks of the Georges, whose shoals of haddock, cod, and hake yearly draw fleets from a dozen nations. East and south, the shallow waters of the continental shelf grade off into the abyssal depths, while on the surface the Gulf Stream brings a breath of the tropics. Though Woods Hole sends oceanographic expeditions to the seven seas, a sizable proportion of its scientists work in their own marine front yard.

The sea, which is Woods Hole's main reason for existence, also helps make that existence pleasant, especially in summer. Fishermen can pursue stripers, tuna, and the more prosaic but delicious flounder; boatmen can anchor in a score of harbors, coves, and inlets. And the water that surrounds the settlement on three sides makes its summer climate salubrious; a forecast of a "hot" day means merely that the temperature may touch 85 degrees. Between June and September a surprising number of scientists or would-be scientists, from grizzled Nobel laureates to undergraduate biology majors, manage to find business that takes them to this small

Massachusetts town.

Research at Woods Hole is carried on by three organizations: the Marine Biological Laboratory (MBL), the Bureau of Commercial Fisheries (a division of the U.S. Department of the Interior), and the Woods Hole Oceanographic Institution. MBL, founded in 1888, is really a collection of laboratories and scientists operating with little or no central direction from the laboratory itself, and much of its work has little, or nothing, to do with the sea. Nobel laureate Albert Szent-Gyorgyi, for intance, has worked there on the function of the thymus gland in mammals. Several other physiologists have been investigating the nervous system of the squid, but their interest is in nerves, not squid; it simply happens that the animal has certain very large nerves that are relatively easy to work with. Nonetheless, MBL has an active program for researching problems in marine systematics (the classification of marine organisms) and ecology (the interrelationships of these organisms with one another and with their environment). Among many other projects, it is currently engaged in taking a comprehensive census of *all* marine organisms, plant or animal, living in Cape Cod Bay, recording their identity, number, and distribution.

The bay interests marine ecologists for several reasons. Cape Cod itself is a "faunal boundary," separating rather different marine communities (the differences depend ultimately on the pattern of water temperatures around the Cape). And the underwater communities in the bay, north of the Cape, are relatively little known, compared to those in Buzzards Bay, Vineyard Sound, and Nantucket Sound to the south. In addition, the bay is part of the near-shore, shallow-water zone that is of peculiar importance in marine ecology because of its role in animal reproduction and its special vulnerability to damage from human activities.

Planning for the census began in 1962 and took three years. The four succeeding years of field work that ended in 1969 involved chiefly the sort of meticulous, repetitive collection of data on which so much of modern science depends.

The four-hundred-odd square miles of Cape Cod Bay were broken up into "quadrats" a mile square, so arranged that by taking samples at each corner of each quadrat, the researchers obtained specimens at one-mile intervals over the entire bay. In each case, work began when the collection vessel—the *A. E. Verrill*—navigated to the center of a quadrat and anchored a radar-reflecting buoy. It then took samples of the bottom, measured water temperatures at several levels, obtained water samples for later salinity measurements, and hauled a plankton net from the bottom to the surface.

Navigating from the buoy, the ship sailed to one corner of the quadrat and sampled the bottom there. Three types of samplers were used. The first was a sort of sled that skimmed along the bottom collecting small fish and crustaceans living within or just over the surface sediments. A modified clam-dredge collected larger bottom-living forms (bigger than two inches) while a scallop dredge was used for intermediate-sized organisms, skimming off the top two or three inches of sediment. On board the ship, the organisms were washed free of sediments (though samples of the latter were kept for reference), anesthetized (to relax them), and preserved in Formalin.

The entire process was then repeated at each of the other three corners, after which the *Verrill* sailed back to the center of the quadrat, picked up the buoy, and moved on to another quadrat. With practice the collecting crew finally managed to cover four quadrats a day, weather permitting. And it took more than a rain squall to stop the *Verrill*, which not infrequently operated in winds of twenty-five to thirty-five knots and seas six to seven feet high.

At the laboratory the sampled organisms were washed again, stored in alcohol, and eventually sorted and classified by genus or family, and in some cases, by species as well. At the same time, the water samples were turned over to the Oceanographic Institute, where their salt content was measured. Later, the samples of sediment were processed to determine the "mix" of sand, silt, and clay particles of various sizes.

These techniques were adequate for most of the bay; in some places, however, glacial boulders on the bottom prevented dragging from above. There, scuba divers collected samples from the bottom.

The bottom-sampling program was supplemented by mid-water trawling for pelagic (nonbottom) fish and other organisms. Fortunately, this sampling did not have to be as intensive, since the upper-water fauna, being generally more mobile, varies less from place to place than the more sedentary bottom-dwelling fauna does.

All the census data is now being classified and coded on punch cards. When run through a computer, the results will enable researchers to make a series of detailed maps showing variations in salinity, in sediment type, and in the identity and numbers of living organisms over the whole expanse of the bay. Because the collections were spaced throughout the year, other, less detailed, maps will show seasonal changes in population. The computer can also determine which organisms, and in what numbers, are found in a given area with which others. Knowing *what* is happening from place to place, it would be very surprising if the researchers could not, at the very least, make some shrewd guesses about *why* it is happening.

It is hard to say whether conclusions drawn from the census will be of "practical" use to anyone. They are part of what is called "basic research," sometimes incorrectly defined as research in which you don't know what the results will be, but more properly, as research in which you don't know whether the results will do anybody any good. Practical or not, however, the whole laborious process of sampling, classifying, and collating minute bits of information, tedious though it may be at times to the participants, is a monument to the lengths to which man, the most inquisitive of the inquisitive tribe of primates, will go to satisfy his curiosity.

In sharp contrast to the emphasis on "basic research" at MBL, the Bureau of Commercial Fisheries is strictly business, and has been since its establishment in 1875. As summed up in a recent annual report, its business is "providing information about current trends in the condition of fish stocks to all segments of the industry," as well as focusing the results of research "on management of fisheries to provide the maximum of benefits to man." A large part of the BCF's resources are therefore necessarily allocated to "the collection and analysis of data on the magnitude and distribution of the commercial catch . . . and on age and length composition of the catch. This provides estimates . . . of rates of growth, mortality and reproduction, all of which are important factors controlling the population density."

As at MBL, "much of the endeavor is tedious and routine," the report notes. BCF agents, located in all major New England ports, collect statistics from fish dealers on the kinds and quantities of fish landed by each boat. In addition, boat captains are interviewed to determine where and how long they fished and their estimated catch in each area.

To improve the accuracy of this information, the bureau is developing a fish-net bathykimograph — a continuous depth recorder housed in a net sinker — which will eventually be distributed to all commercial fishermen. Attached to their nets, it will provide "accurate records of the depth fished and the number, frequency and duration of trawl hauls."

Another research activity of the BCF is the measurement of variations in water temperature, an important factor in the relative abundance of fish. By correlating temperature readings from shore stations, lightships, and its own research cruises, BCF scientists have determined that from around 1940 until about 1953 the waters off New England and the Canadian Maritime Provinces grew distinctly warmer. Since then, the trend has reversed—with some very marked changes in fish distribution: "Shrimp fishermen in the Carolinas are complaining of an invasion of spiny dogfish during the winter months," the report observes, though previously the water was not cold enough for the species. "Sea herring now occur abundantly in the Middle Atlantic area as far south as the Virginia Capes. The codfish population in the Gulf of Maine is on the increase despite increased fishing pressure."

It is tempting to conclude that heavier fishing for cod is nothing to worry about since the fish are now more abundant in the Gulf of Maine than ever before. The real situation, however, seems to be that the population of cod in the northwest Atlantic region as a whole has not risen (and may quite possibly have fallen), though colder waters in the Gulf of Maine have attracted more cod to that particular area.

The over-all abundance of fish depends chiefly upon the proportion of fish eggs that survive to become adults and also, to some extent, on the number of adults laying eggs. But this whole area of recruitment, the bureau notes, is one "in which the greatest ignorance prevails. We do not know when in the early life history of a species the greatest mortality normally occurs, nor much about the causes of this mortality. . . . Is the initial number of eggs important? What are the factors in the environment which are important to survival? How and when do they operate?" To answer these questions, the bureau in 1965 proposed to the International Commission on the North Atlantic Fisheries a large-scale study of the early life history of several species. This would involve collecting eggs and larvae over a wide area and a period of several months, estimating their numbers at each stage of development, and correlating the rate of attrition with such environmental factors as water temperatures and the prevalence of food and predators. The commission approved the idea, and since then, preliminary studies have been going on.

The survey must obviously be a multiship, multination job. A basic requirement will be a standardized collecting device, rugged enough to withstand rough handling and simple enough to be used in all but the worst weather. After a good deal of experimentation the bureau has provisionally settled on a pair of conical nets, with a mesh of about one-half millimeter, towed in tandem to collect fish eggs and larvae, while a similar, smaller pair, with a finer mesh, will simultaneously collect the organisms that fish feed on.

Collecting methods, too, must be standardized, so that specimens obtained by half a dozen nations will all be equally representative of the situation underwater. To test this, the new nets were used on a joint cruise of the BCF research vessel *Albatross IV* and a Soviet research ship, by coincidence also named *Albatross.* Working with the same gear under the same conditions, both vessels, as expected, obtained similar results — but with enough differences, when the vessels were no more than a few miles apart, to suggest that even a small shift in location can mean a sizable alteration in population of marine organisms.

This finding points up the second problem facing the

survey planners: how many samples need to be taken, and at what distances, to obtain meaningful figures? The researchers have reached no final conclusions, but preliminary studies have produced a small practical bonus. Samples taken in an area fifty miles square on Georges Bank, between February and June, revealed that haddock spawning hit a peak in mid-April, when water temperatures had risen from their winter low to between 38° and 42° F. Studies made in the mid-fifties had found that the spawning peak occurred in the same water-temperature range — but a month earlier.*

Woods Hole Oceanographic Institution (WHOI) conducts both basic and practical research. Its work is more diversified than that done by either MBL or BCF, in part because it is much larger. Indeed, its year-round payroll of close to 600 people (750 in summer) makes it the largest private employer on Cape Cod. It maintains a fleet of five research vessels, ranging from 40 feet long to the 213-foot *Chain*, a former Navy salvage ship that has given her name to three ocean-bottom features: Chain Fracture Zone in the Atlantic, Chain Ridge in the Indian Ocean, and Chain Deep in the Red Sea. WHOI also owns a single-engine seaplane, a four-engine DC-4, and the two-man deep-diving submarine *Alvin*, which among other things helped recover the H-bombs "lost" off the Spanish coast in 1966.

The pride of the WHOI fleet is the 210-foot *Atlantis II*, acquired in 1963 at a cost (paid by the National Science Foundation) of nearly four million dollars.

---

* Joint U.S.-Soviet fisheries research is now in its fourth year. The BCF is rather proud of the fact that this is one of the few areas in which the two nations have managed to overcome their mutual suspicions sufficiently to co-operate for mutual benefit.

During her first five years in commission *Atlantis II* traveled some 200,000 miles on thirty-two cruises, including a ten-month trip around the world. WHOI's description of the vessel is an impressive testimony to the complexities and variety of current oceanographic work: "She is the result not only of several years of actual planning and construction, but also of several decades of experience . . . with the seafaring and scientific techniques best suited to a study of the sea . . . the most up to date and versatile deep sea laboratory ever to ply the oceans."

For maximum maneuverability *Atlantis* has extra-large twin rudders and a special bow propeller, mounted crosswise in an underwater tunnel. The propeller can shift the bow to port or starboard or hold it steady against the push of wind and waves. "In addition to the standard operating controls on the bridge, the vessel has four remote control stations located so that the conning officer can coordinate the movements of the ship with the scientific work on deck."

There are "four major scientific laboratories . . . fitted with outlets for special utilities such as fresh water, sea water, electricity, oxygen and other gasses . . . several controlled temperature aquaria which permit the scientific party to keep alive various species of marine life accustomed to either cold or warm water . . . a fully equipped machine shop . . . and an explosives magazine for safe storage of small depth charges used in seismic refraction profiling" (a technique used to plot the subsurface structure of sediment layers on the ocean bottom).

Protruding below the stem is a bulbous underwater

observation chamber with six viewing ports for observing marine life while under way; farther aft is a well that extends through the ship's bottom so that underwater equipment may be lowered into the sea during heavy weather. On deck are two heavy cranes and a heavy-duty winch with six miles of cable. "Underwater lights are located on the skin of the ship near all winches for better night time observation of equipment as it approaches or leaves the surface."

Detailed information on the ship's environment, above and below the surface, is provided by radar, a precision depth recorder, instruments to measure the pull on cables supporting underwater equipment, and hydrophones to pick up underwater noises. The ship's equipment even includes a high-speed computer.

Equipment, however sophisticated, cannot guarantee a successful research trip. Under taxing conditions at sea it often fails to function according to the blueprints—or at all. Two WHOI researchers, John M. Teal and Francis G. Carey, for instance, conducted deepwater experiments to measure the effects of pressure on biological processes and discovered that the pressure also affected their electrical equipment erratically; during the first fifteen minutes under 500 atmospheres the current flow dropped by 30 to 40 per cent. Considerable experimentation was necessary before they managed to devise electrodes that, if not perfectly stable, would at least return to their designed performance level within a few minutes.

Fouling is a nuisance known to boatowner and oceanographer alike. In a survey of the problem another researcher, Harry J. Turner, Jr., found that goose barnacles could foul floating or anchored equipment (such as meters for clocking current velocities) within a few weeks. Ultimately, the rotating propellers that measure current flow would slow down or even stop. Fouling of this sort is a problem only near the surface; lower down, however, equipment may be bitten by fish—at some depths, on an average of eight times a day. In one case the aggressor was identified (from fragments of teeth that it left behind) as an uncommon species hitherto known almost entirely from juvenile specimens. Hoping to obtain a full-grown specimen of this predator (rare enough to have no common name), the researchers set deep lines baited with luminescent lures—but again the unpredictable sea intervened: the gear was lost in a storm.

Particularly instructive, in a depressing sort of way, are the results of WHOI's moored-buoy program. The "buoys" are strings of instruments anchored deep underwater to record data on undersea currents and other physical variables. After an appropriate period a tender homes in on the buoy by tracking the sound emitted by its acoustic "beacon." The tender then sends out a signal of its own, which in theory triggers a release that allows the buoy to rise to the surface and be picked up.

The record of buoys anchored during 1967 reads: "Lost; beacon heard 19 April, not heard 26 April"; "Broke loose; end twisted"; "Anchor line fouled around release"; "Lost after launch"; and so on. Altogether, half the buoys launched were lost in whole or in part, with two-thirds of these being total losses. This rate of recovery, the oceanographers observed sourly, is "un-acceptably low"—especially since the rate did not improve in three years of test launchings.*

Even when underwater apparatus performs according to design, there is still the question of whether it is accurately measuring what it is supposed to measure. The astronomer, though he deals with stars far more distant than the ocean bottom, can at least see what he is dealing with—and can tell you, within a percentage point or so, how many similar stars there are in the sky. The oceanographer, by contrast, collects much of his information blind, at the end of a few score or few hundred fathoms of cable. He is like a man groping at arm's length in a dark closet, from time to time bringing objects into the light for study: he hopes that his finds are reasonably representative of the other objects in the closet—but he is not sure.

The accuracy of this analogy was established by a rather neat experiment devised by WHOI investigator Edward R. Baylor. He coupled a standard type of collecting net with a high-frequency sonar scanner, which could detect animal plankton as they entered the mouth of the net and at any distance up to six feet ahead of it. When the sonar counts were compared with the number of plankton actually netted, it appeared that once an animal had reached the mouth of the net, it was practically certain to be caught. But of the animals six feet in front of the net, only one in six was actually caught when the net was towed at a speed of two knots—

---

* The sea's brutal effects on scientific and other equipment have led oceanographic engineers to speculate wistfully about a mythical construction material called "nonobtanium"—a substance that will do exactly what it is supposed to do under the most adverse conditions.

which is a high speed in comparison with the swimming speeds of plankton. Even when the towing speed was doubled, three specimens escaped for each one caught. Were the escaped animals of the same species as the ones caught? Did each species escape in the same *proportions* as it was caught? The sonar could not answer those questions.

Instead of groping about in his submarine closet, the oceanographer would prefer to walk into it himself, with a light. Vessels like *Alvin*, WHOI's mini-submarine, are beginning to allow him to do just that.

At first glance *Alvin* looks much like a conventional submarine that has been compressed to the size of a smallish truck. It has the usual teardrop-shaped hull, a propeller, and a tiny conning tower. Atypical, however, are the three eyelike viewing ports at its bow and a mechanical grab with which the crew can pick up objects on the ocean floor.

Alvin's most important feature is its cabin—a seven-foot-long hollow sphere of high-tensile steel more than an inch thick. It permits the two- or three-man crew to explore the ocean at depths of up to six thousand feet, where the water pressure is approximately two hundred times the atmosphere's pressure at sea level. "Our scientists," says WHOI proudly, ". . . can view the oceanic environment themselves and selectively pick the samples of rock, sediment or marine life needed for their own particular project or study."

Alvin was first available for intensive scientific purposes in 1967. The reports on its voyages, through the veil of dry scientific prose, give one a hint of the excitement generated by the new scientific tool. One expedition, some five thousand feet down on the continental slope south of Cape Cod, concentrated on observing the bottom or near-bottom fish, sea urchins, and brittle stars—species, the report notes, that "are rarely caught . . . from surface vessels." But on the *Alvin* dives, "we were not only able to observe the animals, but could photograph and capture them as well, even estimate their abundance."

Another expedition exploring much shallower water off Chesapeake Bay, found several ridges, as much as thirty feet high, on the ocean bottom. These were apparently lines of dune-topped sand bars, like those that still fringe much of our southeastern coast, but formed in glacial times, when water levels were lower, and subsequently submerged as the ice sheets melted and the sea rose. "Atop one beach ridge," says the report, "was found a scattered group of oyster shells, the sort of evidence that might be expected to denote the presence of early man." In its submarine wanderings, *Alvin* had apparently happened on an Indian campground of perhaps twelve thousand years ago.

Another shallow-water dive indicated that "a submarine could be used very effectively for taking inventory of . . . commercial groundfish stocks." Still other expeditions explored the underwater canyons that slice through parts of the continental shelf (one of them is now called Alvin Canyon), sampling the rocks along the canyons' sides, layer by layer, with a view to reconstructing their geological history.

Even *Alvin* had its problems, however. One dive had

to be cut short when the craft was attacked by a six-foot swordfish, which managed to wedge its bill into the submarine's outer shell. That night the crew had fresh swordfish steak for dinner.*

Oceanographers, probably more than most research-ers, believe in the scientific maxim known as Murphy's First Law: "If, under a given set of experimental con-ditions, anything can go wrong, it will." Nevertheless, despite aggressive swordfish and clinging barnacles, balky instruments and stormy weather, research pro-jects sometimes go right. The result may be merely another fragment of the incomplete jigsaw puzzle that is oceanography, an unexpected bit of information that upsets some long-held scientific belief, or—once in a long while—a really major scientific discovery whose implications may extend beyond even the rather elastic boundaries of oceanography.

One extensive series of WHOI experiments in the first category has to do with how the color of light changes in sea water. Light passing through any trans-parent medium, such as water or air, is partly absorbed by the molecules it strikes and partly "scattered" by them and by any microscopic particles suspended in the medium. Moreover, the absorption and scattering affects some colors of light more strongly than others.

Looking *up* through clear ocean water, red light is the first to disappear, with the other colors of the spectrum following it one by one until all that is left, at a depth of perhaps one thousand feet, is a pale bluish gray. A

---

* Subsequently the unpredictable sea asserted itself once again: *Alvin* was lost overboard from its tender in rough weather. After more than a year on the bottom, the vessel was recovered with the help of a similar undersea craft, but required extensive reconditioning before it could be put back into service.

different sequence of changes is seen looking *down* at different depths; here the observer sees part of the scattered light itself, some of which is deflected upward whence it came.

The significant point is that the color of the light, whether seen from above or below, is heavily affected by what is in the water—the concentration and type of scattering particles and the abundance of marine organisms, especially plants, most of which possess not only the green pigment chlorophyll but various yellowish, brownish, and reddish pigments as well.

In theory, then, the color of water can give valuable clues to what is in it, and thus contribute to an assessment of its potential productivity. WHOI researchers experimented by hanging radiometers over the side of a boat at various depths and at various locations near Woods Hole and in Massachusetts Bay off Boston. Recorders in the boat drew curves indicating the intensity of light in twenty-five different bands of the spectrum—both "looking upward" and "looking downward." Other "looking downward" experiments were carried out from a plane. By "subtracting" the light directly reflected from the water's surface, it was determined that the air-borne radiometer "saw" much the same colors as had been metered just above the surface.

The investigators reached no startling conclusions. But they pointed out that color differences in water might well be surveyed from the air, or perhaps even from space—thereby delineating different water masses (which at present are identified by extensive and tedious measurements of temperature and salinity) and areas of high or low productivity.

A considerably livelier series of experiments was carried out on seal sounds, by WHOI's William E. Schevill and William Watkins, in the Antarctic and Arctic. In some areas, they reported, underwater seal sounds recorded through hydrophones had "all the variety and noisy flavor of a busy barnyard."

Much of the work was devoted to determining the meaning—to seals—of the various sounds they made. One technique involved recording the sounds and then playing them back through underwater speakers. Provided the fidelity was reasonably good, Schevill and Watkins reported, the animals responded as if another seal were present.

The researchers distinguished three different "meaningful" sounds made by the Weddell seal: trills meaning, approximately, "I'm boss around here"; low pulses, which appear to be a growling threat; and chirps, which may be used to "query" nearby animals, perhaps for navigation, as the similar chirps of the porpoise are used. At any rate, the seals chirped consistently, though intermittently, while making their way from one breathing hole in the ice to another.

The two investigators also recorded a sort of "song" or "trilling call," lasting forty seconds or more and composed of a series of sound pulses. Although they do not say so, it is reminiscent of the porpoise noises described in the preceding chapter. One wonders whether these trills are "communication"—or sonar signals used to range in on fish.

The findings in all these projects are scientifically

useful, and some, such as the water-color studies, may eventually prove practically useful as well. But though they expand man's knowledge of his environment, they do not challenge previously acquired knowledge, and thus are pretty typical of at least 90 per cent of all scientific research.

Among the remaining 10 per cent are the results of WHOI's experiments with the giant bluefin tuna. As popular among gourmets as among sport fishermen, this species is one of the world's largest bony fish (a category that excludes the sharks); its weight can reach nearly a ton. For some years WHOI has, with the help of co-operating sportsmen, been tagging tuna, which are then released to continue about their business. And their business, it now appears, can take them from the tropical waters around the Bahamas to the chilly coast of Norway.

As mentioned earlier, most fish can function only within fairly narrow temperature limits, becoming sluggish or dying in seas much warmer or colder than those to which evolution has adjusted their physiology. The tuna-tagging program, however, has established that the giant bluefin seems to be equally at home off Newfoundland or off Brazil. Not long ago it occurred to WHOI's Francis G. Carey and John M. Teal to wonder how the fish did it. They measured the body temperature of newly caught tuna at a number of locations from Bimini to Nova Scotia, using a thermistor probe—a needle that can be inserted into the animal's muscles to measure the temperature within. The readings showed that the tuna were in all cases warmer—sometimes much warmer—than the waters in which they had been caught.

The coolest tuna, taken in water of about 45° F., had a body temperature of 76°; another, caught under the same temperature conditions, registered 83° F. As the water temperature went up, so did the tuna temperature — but more slowly. In water of 84° F. the fish were only six degrees warmer.

Evidently the bluefin can somehow regulate its body temperature — not as effectively as "warm-blooded" mammals and birds can, but much more so than any other known fish — including its smaller cousins, the skipjack and yellowfin tuna (neither of which is found in water much colder than 70 degrees). The bluefin's accomplishment is all the more remarkable in that, unlike most mammals and birds, it lives immersed in water, a far more efficient body refrigerant than air.

The explanation lies in a physiological mechanism widely found in the animal kingdom: countercurrent flow. It serves a variety of purposes — concentrating salts in the mammalian kidney and in the salt glands of birds, for instance — but in particular keeping body heat in and cold out.

Man's own arms and legs are a good example. Deep in the muscles, the veins and arteries run side by side, but the blood flows in opposite directions. Because of this arrangement, the warm arterial blood, on its way out to the extremities in cold weather, gives up nearly all its heat to the venous blood traveling back to the heart. By the time arterial blood reaches the coldest portions of the limbs, it has little heat left to lose; the venous blood, on the other hand, has been heated to near body temperature by the time it reaches the internal organs whose function would be disrupted by chilling. Cold hands, warm heart, in fact. (In warm weather, when getting rid of heat rather than retaining it is the problem, a system of valves in the blood vessels cuts off man's heat-exchange apparatus.)

It appears that the giant tuna is abundantly supplied with countercurrent heat exchangers in both its gills and its muscles, and they evidently operate efficiently enough for the animal to thrive in places where truly tropical fish would die. The bluefin does not, however, seem to have the "cutoff" system through which man loses heat in hot weather, probably because this is unnecessary to a tuna. Even at the surface, ocean temperatures are seldom higher than 85 degrees, and should a tuna find the heat oppressive, it need only swim down a few fathoms.

Lucky and rare are the scientists who, like Carey and Teal, discover something that, even in a modest way, forces a revision of some long-held scientific belief. Still luckier, and rarer, are those who happen on a really big discovery that is important not merely to one but to several areas of science. Among those fortunate ones are a group of WHOI researchers who in the depths of the Red Sea have found something that is not only new and scientifically interesting in itself but may also constitute an important mineral resource—and throw light on the geological history of central Canada.

The first hint that something strange lay deep in the Red Sea came from a Swedish expedition in 1948, which discovered in several places a layer of bottom water considerably hotter and saltier than anyone had a right to expect even in that sea—which is the hottest and saltiest in the world. They suggested that its source was shallow coastal lagoons where surface water, partially evaporated by the arid heat and weighted by its concentrated load of salt, would flow "downhill" into the deep basins.

This theory was exploded some years later by an English expedition, which hauled up water far hotter and saltier than could be produced by any such natural salt pan. The deep, hot water evidently did not originate at the surface, but rather along the bottom, presumably from the leaching out of salt deposits exposed by fractures in the ocean bed. The resulting brine would soak up heat flowing upward from the earth's interior and, because of its great density, could not dissipate the heat by rising. (Bottom currents could not help much either, since the Red Sea's bottom waters are cut off from those of the Indian Ocean by a relatively shallow sill across the strait of Bab el Mandeb.)

This explanation was consistent with a number of other features known, or suspected, about the Red Sea. It is part of an extensive system of "rifts" in the earth's crust—narrow belts in which the crust is being pushed upward and thrust apart. Rifts are found in all the oceans and, in places, on the land as well. (The Red Sea portion, in fact, continues several hundred miles north through the valleys of the Dead Sea and Jordan River; its gradual opening, over the next million or two years, will thrust an arm of the sea between Israel and Jordan.)

The undersea-rift theory neatly accounted for the Red Sea's deep, hot brine pools: the rift naturally produces cracks through which the salt deposits, thought to be present in that region, could leach out. Rift zones, moreover, are regions of abnormally high heat-flow.

In 1965, and again in 1966, WHOI expeditions cruised the Red Sea, mapping and sampling the ultra-briny deeps and their underlying sediments. Three such areas were found—Discovery Deep (named for the British oceanographic vessel that had spotted the phenomenon some years before), Chain Deep, and largest of all, Atlantis II Deep, an irregular trough some eight miles long and three miles wide. Atlantis II is now thought to be the source of all the hot brines; from time to time these become abundant enough to flow over the seabottom "saddle" that separates that deep from the other two salt pools.

The WHOI researchers found water hotter and saltier than even the British had discovered. Temperatures exceeded 130° F. (about that of a July day in Death Valley, and some forty degrees above the maximum Red Sea surface temperatures); and salinity was an almost unbelievable 27 per cent—about eight times the usual salt content of ocean waters.

Most remarkable of all, however, were the bottom sediments in Atlantis II Deep. The uppermost sixty feet is a thick mud that is unexpectedly rich in compounds of copper and zinc, with smaller amounts of silver and gold. It appears that the deep is a sort of undersea pressure retort, in which minerals from the surrounding rock are first dissolved, then concentrated, and finally precipitated out to sink to the bottom.

These sediments were assayed at nearly three dollars

a cubic foot for the bulk mud; the entire deposit may be worth several billion dollars. As commercial mineral deposits go, the Red Sea muds are not extraordinarily rich—but may be rich enough to be "mined" profitably, if methods can be found for dredging them up.

With an ore deposit of this size in an undersea rift, it is natural to wonder whether other deposits, on land, might not have been formed in the same sort of submarine pressure cooker. Hardly had the Red Sea findings been published than evidence began coming in to suggest that this may in fact have happened. E. R. Kanasewich of the University of Alberta, using extensive geological studies of south-central Canada, has mapped what appears to be just such a prehistoric rift, formed nearly a billion and a half years ago and long since buried by other sediments. It stretches through parts of Alberta and British Columbia, and perhaps across the border into Idaho and Montana; within it lie the rich Kimberly lead and zinc mines, and perhaps also the silver, copper, and zinc deposits of Coeur d'Alene, Idaho.

That mud dredged up from six thousand feet beneath the Red Sea should throw light on a geographic puzzle in Canada, half a world away and some hundreds of miles from the ocean, is less remarkable when we recall that central Canada, like almost every other spot of what is now dry land, was at one time or another part of the ocean bottom. For we must end this book as we began it—with the sea's dominance over the land, eroding it and replacing it, intruding on it and drawing back, its moisture essential to life, its riches essential to man and his civilization. Our species has lived on land for tens of thousands of years, and our ancestors for hundreds of millions before that, but our habitat is still locked in the ocean's encircling embrace — and to understand the one, we must also strive to understand the other.

Having said this, I must add one thing more: understanding alone is not enough. Research centers such as Woods Hole are providing us with an ever more detailed picture of the physical and biological processes that make the ocean what it is; they are tracing the intricate web of relationships that connects the sea's living organisms; they are discovering the ways in which human activities are changing, and can change, those relationships. Research can tell us, in short, not only what the sea does to and for man but what man is doing, and might do, to and for the sea. But all this human effort — its imaginative theories, its intricate technologies, its stultifying yet essential piling up of sand grains of information — will be but a sterile exercise unless men, and governments, are willing to act on the knowledge that the scientists have laboriously won. That willingness, or the lack of it, will determine whether the sea will remain what it has been in the past, an ever-renewed source of riches, or will become (in the words of S. J. Holt, of the UN's Food and Agriculture Organization) "a contaminated wilderness or a battlefield for ever sharper clashes between nations and between the different users of its resources."

On sea, as on land, science can only point out the consequences of our actions; how we then choose to act is up to us.